AF335554

Drawing of the 200-inch Hale telescope by Russell W. Porter
Hale Observatories photograph

RUSSELL W. PORTER

Arctic Explorer • Artist • Telescope Maker •

by

BERTON C. WILLARD

With a Preface by David O. Woodbury

The Bond Wheelwright Company
Publishers • Freeport, Maine

PICTURE CREDITS

Illustrations courtesy of the following:

Mr. Fred F. Chellis: page 158

Mr. E. V. Flanders: pages 163, 256

The Hale Observatories: frontispiece; pages 191, 192, 196

Mrs. Caroline Porter Kier: pages xiv, 2, 4, 6, 10, 16, 27 (water color); 34, 36, 40, 42, 43, 47, 51, 58, 66, 68, 72, 74, 76, 82, 92, 98, 101, 103, 104; bottom page 106; 107, 109, 111, 112, 118, 124, 130, 176, 182, 183, 184, 194, 203, 204, 206; bottom page 207; 227; bottom page 228; 242, 244

Mr. Lafayette Marsh: pages 26, 84

Mr. Alan Rohwer: pages 100, 164

The *Scientific American*: pages 136, 152, 153

Springfield Telescope Makers: pages 15, 27 (photo); 28, 29, 30, 31, 122, 132, 134, 138, 139, 140, 142, 144, 146, 151, 152, 153, 162, 163, 166, 167, 168, 169, 186, 188, 189, 195, 210, 211, 214, 215, 216; bottom page 218; 220, 222, 224; top page 228; 230, 231, 234, 258

Mrs. Helen Perry Smith: top page 218; pages 226, 237

Dr. Bradford Washburn, Boston Museum of Science: page 93

All of the paintings and drawings and many of the photographs are reproductions made by the author for this volume.

Library of Congress Catalog Card Number 76–8090
ISBN 0–87027–168–7 Hardbound
ISBN 0–87027–167–9 Paperbound
In Canada: Beaver Books, Fitzhenry & Whiteside, 150 Lesmill Rd.,
Don Mills, Ontario M3B 2T5
Composition by Pine Tree Composition, Lewiston, Maine
Manufactured by Murray Printing Company, Westford, Massachusetts
First printing

To my wife, Sylvia, whose loving patience, understanding, and help made this book possible.

CONTENTS
AND LIST OF ILLUSTRATIONS

ILLUSTRATIONS

"Something hidden. Go and find it. Go and look behind the
          Ranges—
"Something lost behind the Ranges.
          Lost and waiting for you. Go!"

—Rudyard Kipling

This excerpt from "The Explorer" was Russell Porter's favorite
quotation and typifies his innate lifelong desire to find out for
himself.

To promote the ideal that one should lead an adventurous
and creative life, however great or small, has this book been
written.

Berton Willard

PREFACE

When Russell Porter died in Pasadena, California, in February, 1949, at the age of 77, the world of American science lost one of its most brilliant yet unassuming workers—a man, gentle but vigorous, firm but soft-spoken, lovable but adamant and courageous in his principles—altogether the model for scientific integrity, while yearning always to push ahead into new concepts and new ground.

Because of his affectionate appeal, Russell had many thousands of friends and admirers, not scientists in the main but taken from the wide and earnest amateur fringe of science. People everywhere loved him because he was a genuine, no-nonsense experimenter and pioneer. His field was astronomy but his appeal was that he could, invariably, make the most abstruse concepts seem simple. He symbolized the will to invent, to take chances for great gains, and he made every one of his followers somehow understand and cluster around him. Russell was a true leader of thought and performance. He opened constructive astronomy and gave it as a gift to everybody within his reach.

It was a unique and difficult combination which this modest man commanded, but it was a talent which now lives on in these pages by the able hand of Bert Willard, his only biographer. We have here a chronicler who, though he did not know Porter personally, has had full access to his subject's voluminous diaries and innumerable letters, articles, and accounts, which he and his intimates provided through the years. As a late-coming member of the Stellafane group and a student of ATM, which Porter and Ingalls had written, Willard is in a splendid position to give a careful and authentic history of Porter's life and work.

When I arrived in Pasadena in 1938, to write my *Glass Giant of Palomar*, Russell was at the door of his home on California Street to greet me, a stranger. From that moment on he never flagged in his joyful resolve to exhibit for me the magnificence of the 200-inch concept. From the start, he managed to make it all seem possible of accomplishment, though at that time there were still most serious engineering and even human problems in the project. Russell knew every angle of the great saga, and why not? By that time in his life he was part and parcel of the most noteworthy story the world had in astronomy. He knew the Cal Tech and Mount Wilson personalities intimately—information I had to have if a book was to be wrested from them. And he knew instinctively which of these prima donnas it would be wise for an outsider to avoid. Only Hale was missing; he had died.

Through the wisdom and friendly help of Russell Porter, we discerned that I was suspect in the Observatory Council, who, confronted by a mere author, looked down its nose at me and remarked severely, "Let him begin, then, by making a telescope himself!" How else? smiled my mentor, Porter. Had he not come to Pasadena himself by somewhat the same route? A route, by the way, quite unfamiliar to publishers, one of whom had told me in New York, "Go get the story, do the book." I could not help wondering in those puzzling days how Porter, himself, had scaled this fearsome astronomical wall, even at the request of Hale himself. But he seemed always to have belonged, always to have skillfully circumnavigated the doubts and prejudices with which scientists of high order invariably crush down outsiders. Somehow, Porter had sailed in and not only conquered, but become an inspiration to them all.

This inspirational talent of Porter was what I knew best, because he was unfailingly willing to go to any trouble to explain the mysteries obscure to me. And his dear wife Alice was just like him. She knew, instinctively, how to make his talent shine most brightly. And I think our author, too, has caught the spirit, through his later experience at Stellafane.

One of the choicest moments I remember was when my wife and I got "Russ and Alie" to visit us in Ogunquit, Maine, where my father was long established teaching painting. Father had not seen hair nor hide of Russell since old MIT days so many decades before, when he had taught him to paint. Yet now they put their heads together and reminisced the years away as old cronies should. Russell, as we had long known, was a genuine and skillful artist now. Some of his early arctic things were magnificent; later, the Alaskan wilderness partook of his talent likewise. Of course, his unmatchable cutaway drawings had started a new technique in art altogether. All this and more, made my father very pleased indeed. Russell was no amateur—*in anything!*

While Mr. Willard's book is not mainly a chronicle of the Palomar days, he has blended them smoothly into the earlier Stellafane experiences that certainly made Porter eligible to take an official position of real importance in the 200-inch telescope effort. The author gives every indication of understanding the metier in which Porter and his disciples have all worked—that mystical light from the heavens that trickles through the souls of the fashioners of glass. I am glad that he offered his time and skill for the task, and I wish him every success with the result.

David O. Woodbury

The earliest known photograph of Russell W. Porter
  Date unknown

I. EARLY YEARS AND THE ARCTIC

The village of Springfield is wedged between steep green Vermont hills, and the business center must share what little level land there is with the Black River. Long before any man knew the surrounding hills, the river had worn itself into a deep gorge. It was this source of easily tamed water power that had attracted the settlers to the valley in the first place.

Young Russell Porter approached the bridge as he had done countless times before walking between his home on Park Street Hill and the center of the village. But this time the boy had made up his mind. He would do it now. Cautiously at first, perhaps to see whether he was going to change his mind, he walked to the end of the bridge and looked down at its iron structure, typical of the 1880's construction. Fifty feet below, the river was squeezed to a swift current by the narrow gorge. The water rushed around boulders, leapt over rocks, and churned into white madness as it spilled down over the falls.

Russell left the safety of the roadway and took hold of the iron struts. Keeping a firm grip and a sure foot he climbed down until the dark underside of the bridge was above his head. He was in the gorge. The wind blew between the sheer walls, cool and moist against his back and arms. Only his hands felt the cold iron. From the corner of his eye, he saw a neighbor standing at the edge of the bridge staring down on him without a word. In a moment he forgot the figure as he concentrated on moving farther out along the frame-

*Russell Porter's parents, Caroline Silsby and Frederick Wardsworth Porter
Date unknown*

work. Finally he reached his goal at the center of the bridge. The deadly torrent was directly below him. He never saw the group of villagers who were gathering on the bridge. No one shouted for fear of distracting him. Then, his exploration ended, he worked his way back to the side of the bridge and on up to the road and safety.

"I wanted to see what it was like" was his only explanation to the astonished and relieved crowd.

Walking home up the hill, his curiosity satisfied, he had learned at an early age to face danger with confidence and a will to overcome. He would come to rely on this lesson during his years of arctic exploring.

Frederick Wardsworth Porter was a prominent citizen of Springfield by the time his son, Russell Williams, arrived in the family on December 13, 1871. He was an industrious man who had developed a mechanical inclination at an early age. With the guidance and tools of an uncle he constructed a small steam engine when he was eleven and used it to drive a small boat on the river. In 1841, at eighteen, he decided to learn the art of taking pictures by the Daguerrean process, first made public in 1839, and took instruction from a New York photographer. The daguerreotype was a silver-coated copper sheet which was developed by mercury vapors. Frederick was successful enough to make it a profitable enterprise for two years before he decided to go into the jewelry business as his permanent vocation.

In 1845 Porter opened a jewelry store in Springfield that he ran for thirty years, adding books, stationery, and drugs along the way. Watch repairing helped to satisfy his desire to tinker with mechanical things. Tiring of all this, he later formed a partnership and set up a toy manufacturing company which prospered for many years. The toy carriages the firm produced, sometimes as many as 75,000 a year, were sold throughout the country.

He always bore the appearance of a polished gentleman and was of a retiring disposition. The townspeople respected him, and he held numerous public offices, among them town representative, school district official, and postmaster.

Porter's wife, Caroline Silsby Porter, was born in the village of Charlestown, New Hampshire, just across the Connecticut River

The Porter home and Russell's birthplace in Springfield, Vermont

from Springfield, Vermont. Both were religious followers of Emanuel Swedenborg, the eighteenth-century Swedish scientist, philosopher, and theologian who is considered one of the greatest and most learned men of his country.

Russell was the last of five children. His father was forty-eight and his mother was forty-five at the time of his birth. The Porter children were widely spread in age; the oldest of Russell's brothers and sisters, Anna Silsby, preceded him by twenty years. Then came William Bradley, sixteen years his elder, Frank Farrington, older by twelve years, and Elizabeth West, almost three and a half on December 13, 1871.

The Porter home was located a short walk from the center of the village, across the Falls Bridge. Part way up curving, tree-shaded Park Hill and on the left was the large brick and wood house that became known as "The Evergreens." A garden with trees, shrubs, and flowers attracted much local attention. In the back a barn kept the Porter livery.

Few facts have survived about Russell's youth. He attended the public schools in Springfield, where his friends thought him lazy. He had black hair and blue eyes and a round face that could radiate determination. He would mature to a five-foot eight-inch frame of about 150 pounds.

Like any normal boy he was not always angelic. One day his father discovered some broken window panes in the barn gable. Tracing the damage to his son, the elder Porter gathered up some pebbles, handed them to Russell, and instructed him to throw the pebbles at his face. That put a quick end to his glass devilment, although it did not prevent him from later attacking glass with a scientific purpose.

In 1887 Porter went to Vermont Academy in Saxtons River, a few miles south of Springfield. There he studied for two years in preparation for college. In September 1889 he entered the sophomore class of Norwich University in Northfield, Vermont. He became a member of the Theta Chi Fraternity. Then, in the fall of 1890 he transferred to the University of Vermont at Burlington. Here he settled down to studying civil engineering.

Soon Russell's plans were shattered. Suffering from large economic losses, his father discovered he could no longer afford to send

The earliest surviving drawings by Russell Porter, done at the age of 15

his son to college. With no money of his own, Porter was forced to leave school and seek employment in 1891.

Feeling that the chances of success were better in the city he traveled to Boston. He found a job with an insurance company appraising buildings, which gave him an opportunity to examine their construction in detail. Fascinated by the design of these buildings, he decided to become an architect. Surprised to find that his employer encouraged his dream, Porter enrolled in the architectural course at the Massachusetts Institute of Technology in the fall of 1892. By continuing a part-time appraising job with the insurance company and borrowing money, he was able to make his own way through the first year at MIT. A successful career finally seemed assured.

But the architectural dream of that first year was not completely unscathed. In the fall of 1892 at the age of twenty, Porter attended a lecture on arctic exploration. Robert E. Peary, who had recently completed his first crossing of northern Greenland, was raising funds

Porter as a young student at MIT

to continue his work in the North. "Out of a clear sky that evening came to me the undefinable lure known as arctic fever," wrote Porter. For days afterward he could see nothing but Peary's pictures of "snowdriven figures toiling over the endless wastes of Greenland's ice cap, drawn towards that point where all longitudes meet." [1] He had to go and see for himself. He had to look behind the ranges.

He talked to his friends about his desire to go north, but they tried to discourage him. He talked with his parents, but of course they were opposed to the idea. During the winter he went to Washington to have a personal interview with Commander Peary, hoping to be taken on his 1893 expedition. Peary agreed to consider Porter but later wrote that he could not accept him. Years later Porter learned that his mother, then sixty-seven, had written to Peary begging him not to take her son into the Arctic. Thus ended his first attempt to meet the white goddess who seemed to be beckoning to him. He was disappointed but he did not give up.

During his second year at MIT, in 1894, another lecturer on arctic exploration came to Boston. This time it was Dr. Frederick A. Cook, who had been with Peary in Greenland. Wanting to try something new in arctic travel, he was advertising a summer cruise up the west coast of Greenland in a chartered vessel, the cost to be $500.00. Porter was able to talk Cook down to $300.00 by agreeing to be the artist and surveyor for the party.

The surveying he had learned during his studies at MIT. Of Porter's first signs of artistic ability there is no record, but he was already skilled with a pencil and brush by the time he went to Boston. There he had met the artist Charles H. Woodbury, who was a member of the MIT class of 1886. Woodbury, destined to be one of the foremost American marine painters, was in Boston for several years teaching painting, and Porter took advantage of the opportunity to study with him.

Porter was also working his way through school as a draftsman in the architectural firm of Codman and Despradelle. With a full beard and many French architectural honors to his credit, Despradelle was also one of his professors at MIT. Now in the summer of 1894 Porter took a leave from the drafting job and headed for his first arctic voyage. There were many later times when he left the firm for the summer and many times when Despradelle welcomed him back, always with the hope that he would at last settle down to an architectural career.

In response to Dr. Cook's advertisement, fifty-two men and a thirteen-year-old boy joined him on the *Miranda* in New York Harbor. The men came from many professions—university professors, doctors, surveyors, photographers, sportsmen—and all age groups. They finally got under way on July 7, 1894, anticipating an exciting summer of studying Greenland's glaciers, icecap, and Eskimos, exploring and mapping the coast of Melville Bay, and hunting. But the excitement they found was unexpected.

The *Miranda*, an iron steamship, was unsuitable for arctic travel. At every port where she stopped from New York to Greenland, the men were met with dire warnings of their pending doom in the iceberg-laden sea. She came to the cruise with a long history of mishaps, among them running on the rocks, sinking and being raised, colliding with a steamer and a schooner, and finally being forced from passenger service to cargo hauling. Those who had never experienced her antics were soon initiated. Instead of backing out from Pier 6 in the North River, the *Miranda* rammed the dock. Somehow the command came to the engine room reversed, probably the fault of newly repaired signal wires. Then, early one morning in thick fog off the straits of Belle Isle, when all seemed well at the breakfast table, a commotion was heard on deck. Suddenly a deafening crack and then a convulsive shudder ran through the ship. The men raced up the companionway to come face to face with a towering iceberg. Blocks of ice were strewn about the forward deck. The iceberg had done its damage and the starboard bow-plates were stove in, but fortunately the hole was above water line. Now it was necessary to head for Cape Charles, Labrador, and then to St. Johns, Newfoundland, for repairs. Some of the passengers felt, perhaps rightly, that if this was any indication of what lay ahead they wanted no part of it. Hunting in Labrador was not a bad second choice to Greenland anyway. Porter disagreed. Unruffled by the mishap, he was as determined as ever to get to Greenland.

Finally the *Miranda* was made seaworthy, and most of the original party again set forth for Greenland, braving fog, floe ice, and more icebergs until landfall was made on August 7 at Sukkertoppen, Greenland, a Danish-governed Eskimo settlement. There

Porter drawing of the Miranda *after collision with an iceberg off the*
Straits of Belle Isle while bound for Greenland in 1894

they planned to stay only a short time and then continue up the coast before the season became too far advanced.

For Porter the excitement of finally being in the arctic region quickly compensated for the disappointing months since he first heard Peary lecture in Boston. Although they were still a little south of the Arctic Circle, the scenery was magnificent. Porter photographed rolling hills covered with tundra, distant mountains barren of trees, and fjords snaking through the mountains, but unfortunately the film never reached home. "I love this land," he wrote in a letter to his mother, which he sent by kayaker and ship to Copenhagen to be mailed. "How different it all is from what you and most people imagine. I knew it was so and find [it] even more charming than I had supposed." [2]

Most of the party, however, never got farther north than this little town. Shortly after leaving the harbor on the morning of August 9, the *Miranda* moved over a submerged uncharted reef. As she passed into deeper water, the sound of tearing metal could be heard throughout the ship. The panic soon subsided when it was discovered that only the after water-ballast tank was full of water.

Years later Porter wrote: "I had for the first time been thoroughly frightened, and tasted what, I think Stewart Edward White calls 'copper.'" He had this taste several times during his life, "at times of great fear when events were transpiring the outcome of which seemed fatal to me." It was a definite taste, "akin to the sensation one feels when placing upon the tongue certain electrodes that will generate a slight current." [3]

The *Miranda* was able to return to the harbor, but she was decidedly not safe enough for the trip home without an escort. Unless a fishing schooner could be found off the coast, an unexpected winter in Greenland was a definite possibility. There was little to cheer in this prospect because Sukkertoppen could not possibly support the population of the *Miranda*. So in an open sailboat lent by the governor, Dr. Cook commanded a party of five of his men, one of them Porter, together with a Dane and four Eskimos as guides. They sailed to Holsteinborg, about 150 miles north, in search of American fishing schooners known to be in the area.

On this trip Porter first crossed the Arctic Circle. "I doubt whether the others experienced the thrill of passing this imaginary line N. 66–⅔ degrees, dividing the temperate from the frigid zones of our earth, but I had scanned that dotted line on maps too often not to be enthused." [4] Just beyond the circle, at the town of Holsteinborg, the men learned from the governor that several fishing schooners were somewhere off the coast not far away. An Eskimo kayaker carried the distress message to the nearest, the *Rigel*, five months out of Gloucester, Massachusetts, and her captain and four fishermen returned to Holsteinborg in a dory.

After hearing Cook's story and his request for help, Captain Dixon said he could not forsake his fishing without first conferring with his crew since they were fishing on shares. He returned to the *Rigel* with the understanding that if she appeared in the harbor the following morning the crew had agreed to come to the rescue of the stricken *Miranda*. Early the next morning anxious eyes began scanning the sea. Finally the *Rigel* appeared. They would not have to winter in Greenland. The captain returned Cook's small party to Sukkertoppen on August 20.

The plan was for all passengers to squeeze into the *Rigel* as best they could among the fish, salt, and gear while the *Miranda*, her crew aboard, towed the *Rigel* back to Newfoundland. But on the third day out, the *Miranda* had to be abandoned to the sea; her ballast tanks had given way under a pressure they were not designed to take. After the *Miranda's* crew was transferred to the *Rigel*, her hawser was cast off and she steamed away into the fog as if determined to cruise forever. No human eyes saw her go down for the last time.

Now with ninety-three men crowded onto the *Rigel* and provisions running low, seriously hampered by fog, icebergs, high seas, and head-winds, they made what haste they could toward the Labrador coast. On the evening of August 28 they put in at a small harbor in Labrador and then continuing south among icebergs shrouded by fog entered Sydney Harbor, Cape Breton Island, safe at last on September 5.

At this port some passengers chose to return to New York by land. Most, including Porter, continued by the steamer *Portia*, arriving in New York on September 11. Bad luck had followed them, however, for now, entering Long Island Sound in fog, the *Portia* ran down the coal-laden schooner *Dora M. French* and sent her and most of her crew to the bottom.

Most of the men who had signed on for the *Miranda* voyage probably never again saw Greenland or the Arctic. The one experience had been enough to satisfy many adventuring minds. But one mind was not satisfied. To Porter the "fugitive glimpses of the shimmering ice cap" of Greenland, "the haunting melodies of the dancing Eskimos" and the first crossing of the Arctic Circle only whetted his appetite for that great mystery, the North. His "first flirtation with that icy goddess residing beyond the Circle" [5] had ended in failure, and he barely made it home with the clothes on his back. Even so, the trip had raised his "arctic fever," the term he used again and again to describe his attraction to the North.

In the fall Porter returned to Boston, took his seat in the academic community, and resumed his part-time job with Professor Despradelle. The architect assumed that the wanderlust was satisfied and Porter was ready to settle down. However, he knew nothing of the impressions that had been planted in the mind of this man. "As I leaned over the drawing board I could hear the ice pans grinding together, the raucous cries of the gulls and the gutteral ejaculations of the Eskimos." [6] Porter, struggling to keep his mind on his studies and work, was unable to forget his summer experience.

During the fall and winter of 1894–1895 Porter roomed in Watertown and each morning walked into Boston to MIT. He often met and walked in with Alfred Shaw, a classmate who lived in Newton Highlands. Shaw was as much fired up for adventure as was Porter, and the two young men spent many hours together. On weekends they canoed and camped on the Charles River. Once they discovered that with a few short portages they were able to circumnavigate the city of Boston. Other hours they spent planning expeditions to countless remote areas across the globe.

The following summer, 1895, was a total disappointment. The two adventurers, unable to locate a single arctic expedition, made plans to cross the Atlantic by cattleboat for a walking trip through England. Then Professor A. E. Burton told Porter that Peary was going north in 1896 and had asked Burton to be leader of a party of scientists to be landed somewhere along the west coast of Greenland. Could Porter be talked into going with him? No talking was necessary; Porter would go in any capacity. The trip to England was cancelled. Probably Porter felt it prudent to work for Despradelle this summer and be in a better position for an important arctic trip the following summer. Thus, another winter at MIT passed before he could meet with his arctic goddess.

But in 1896 Porter began his second arctic voyage, this time on the whaler *Hope,* which was carrying Peary and his party to north-west Greenland in search of a large meteorite. The scientific party of about six men under the leadership of Professor Burton was put ashore at Umanak, a little Danish settlement well above the Arctic Circle. Porter's responsibility was the commissariat, the food and camping equipment of the party staying in the big Umanak Fjord.

Umanak was presided over by Governor Knudsen, a Copenhagen University man. He and his wife had little contact with the outside world and could not do enough for the men. Porter was given a small house of his own, which he turned into a studio. Earlier, in 1869, the American painter William Bradford, 1823–1892, had visited the Greenland coast. Porter greatly admired his arctic paintings and was perhaps inspired by them to try his own hand at painting in the Arctic. In his studio he produced full-sized water color portraits of the natives, which still stand among the finest ever made of the Greenland Eskimos.

"The women posed in their best skins," Porter wrote "with colored ribbons tied tight about an upstanding topnot (*sic*), the particular color worn denoting their social standing in the community." Red showed the woman was unmarried and eligible, blue that she was married, and black that she was a widow. Sometimes there was "a little white mixed in with the black, . . . these ladies who had lost their husbands, were still in the market for others." And, Porter discovered, the "badge of disgrace was green—neither maid, wife nor widow." [7]

One day Porter made a water color of the governor's house and presented it to Mrs. Knudsen. Six years later, returning from Franz Josef Land with the Baldwin-Ziegler Expedition, Porter went to Jutland to visit Mr. and Mrs. Knudsen, who were retired and living on a farm. They showed him on their bedroom wall his water color of their home in Greenland.

Porter not only painted; he helped the scientists with their work. Because they were interested in measuring the movement of the Karajak Glacier at the upper end of the Umanak Fjord, Porter had a chance to get a closer look at a glacier and the inland ice that he

Porter water color of an Eskimo girl, done in Umanak, Greenland, 1896

 Russell W. Porter

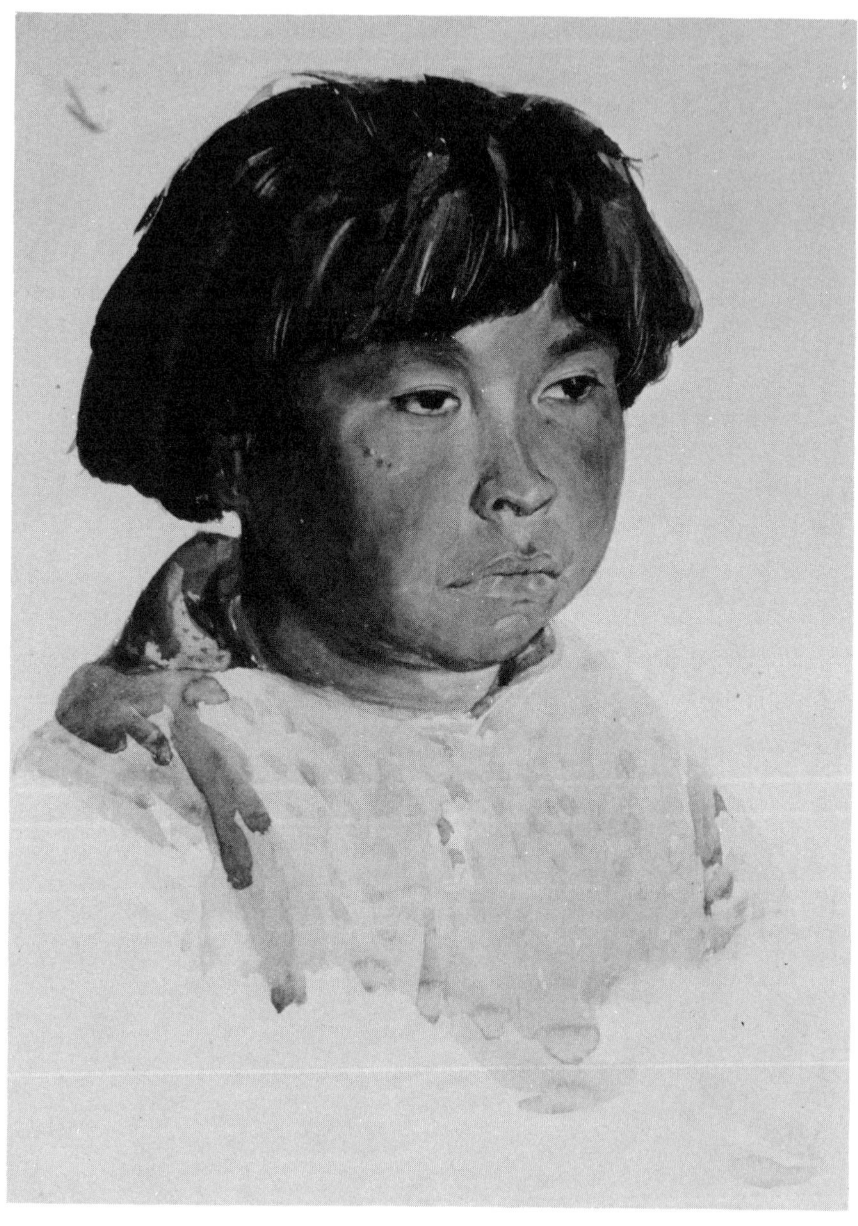

Porter water color of an Eskimo boy, done in Umanak, Greenland, 1896

had only glimpsed two years earlier. The party moved up the fjord in two boats heavily loaded with supplies, white men, Eskimos, dogs, and sleds. The fjord was full of infant icebergs of glacial parentage, which had to be avoided constantly.

Near the glacier's end the party made camp and set up a line of stakes on its surface. They followed the glacier up onto the icecap until all sight of land had vanished and all that was visible was an unending expanse of snow and ice in every direction. Returning to the coast, they measured how far their stakes had moved downstream on the glacier. During the trip Porter and another man tried their luck at hunting caribou but without success.

Now Peary returned from farther north and gathered up the scientific party. He had not been able to retrieve the meteorite. (This failure would prove to be a stroke of luck for Porter the following year.) After crossing Baffin Bay on the way home Peary put in at Cumberland Sound on Baffin Island so the geodesist could once again swing his pendulum to determine the gravitational constant. Porter used the time to talk with a Scotsman who was running a whaling station. He learned that most of vast Baffin Island was still unknown to white men, a blank on the maps. He learned of native reports that there were lakes in the interior so large that they could not be seen across. There was even the possibility of tribes living there completely unknown to the outside world. That was all that Porter needed. Immediately plans began to form in his mind: he would organize a small expedition of his own and return the following summer to become the first to explore this unknown land.

But now he had to return to Boston and once again take up his drafting job. Once again Despradelle welcomed him back. Would the wanderer now settle down and work toward his career? At the same time Porter received a scholarship from MIT and spent the school year 1896–1897 taking postgraduate courses even though he had not completed the undergraduate program. He never earned a degree from any school.

Headwaters of Frobisher Bay, Baffin Island, 1897
Porter water color

II. FOOTLOOSE AND ARCTIC BOUND

During the winter of 1896–1897 Porter found it difficult to concentrate on his studies and his work with Despradelle. How could he become the first explorer to penetrate Baffin Island? How could his dream be financed? In his spare time Porter searched for a backer, someone to provide the necessary money to outfit his expedition. Finally, in the spring of 1897, after a fruitless winter of searching, he received a letter from Peary.

Peary was going north again that summer to make another try for the meteorite and invited Porter to organize a group of students to spend the summer in Frobisher Bay on Baffin Island. He suggested that the students return with him and Porter remain over for the winter. This way Porter could carry out his serious exploring with a single companion and be taken home during the following summer. Here was his chance.

With some doubts about his ability to organize a party of students, Porter advertised in the college papers. For $500.00 apiece he would provide passage to and from Baffin Island and all necessary equipment for two months of hunting and fishing in Frobisher Bay. It worked. Soon he had six students eager for just such a summer's outing. He even arranged with the Scotsman at the whaling station at Cumberland Sound to give him passage on his ship to Dundee, Scotland, the following summer (although the arrangement was not in the end used). His friend Alfred Shaw planned to winter over with him.

Peary's ship *Hope* left Boston on July 19, 1897. As they traveled northward Porter could not help but reflect on the history of Frobisher Bay. The bay was discovered by Martin Frobisher in 1576 during an unsuccessful attempt to find the northwest passage. Frobisher went on to serve Queen Elizabeth and gain distinction in the British Navy on various expeditions. When the Spanish Armada threatened the island nation, Frobisher was among the four men, including Sir Francis Drake, most instrumental in repulsing the attack. He was knighted for his valor in this command.

Soon after 1576 a group of English colonists sailed into the bay in search of gold reportedly discovered by Frobisher. Some were never heard from again, and none brought back gold. Then in 1861 Charles F. Hall reached the headwaters of the bay, proving it was not the northwest passage. He also found traces of the Frobisher adventurers on Kodlunarn Island in Bear Sound near the bay's entrance. There was no written evidence that white men had visited the bay since Hall. Porter hoped to find further traces of the gold-seekers of over three hundred years ago. But his main intention was to be the first white man to explore the interior of Baffin Island and make a study of the Eskimos.

On August 1, the *Hope* landed the Porter party at a whaling station run by a Dane on Cape Haven near the entrance to Frobisher Bay. Peary then continued north, leaving the little band to its own resources for the summer. "We must have presented a strange appearance," Porter wrote, "as our fleet of two whaleboats, loaded to the gunwale, debouched into the bay for the summer's cruise." Three or four Eskimo men had been persuaded to guide the explorers "for the princely pay of one plug of tobacco a week each." [1] Of course the Eskimos threw into the bargain their entire families and all their worldly belongings, women, children, babies, dogs, and puppies.

The white men soon picked up enough Eskimo words to converse. One day Porter overheard one of the Eskimos use the word "Kodlunarn," referring to the "White Man's Island" in Bear Sound explored by Hall. The Eskimos took them to the island, and there Porter and Shaw camped one night and made a thorough search for evidence of early white habitation but without success. Thus ended their search for the Frobisher colonists.

The bay was found to be about 140 miles to its head and thickly strewn with islands. The party was fortunate to have the natives as pilots through the tricky tidal currents. The Eskimos seemed to possess uncanny knowledge of winds, weather, and tides and skillfully piloted the boats around the worst currents. At the head of the bay the tides rose and fell forty feet. Here, near the mouth of the Jordan River, camp was made, for the Eskimos were eager to join some of their tribesmen whom they had not seen since the previous winter.

It was a land of contrasts. A desolate landscape first met the eye, yet the area was a horn of plenty. Where the river tumbled into the bay, the Eskimos speared sea trout. The hunters of the party were pleased to find caribou in the area.

The rocky land was slow to rise from the bay. When it did roll into low hills, they everywhere seemed the same, never rising as high as 2,000 feet. The tundra gave little encouragement to growth. It managed to survive only in the shallow valleys and hollows, protected from the blast of arctic storms. Tiny streams wound through the rocks and tundra, fed by rain and melting snow. Rocky ridges meandered over the landscape, joined with other ridges and disappeared over the horizon.

Near the camp they visited Silliman's Fossil Mount, a limestone table-topped hill, abundant in fossils. Since no one in the group was a paleontologist they collected only the most attractive specimens. Later, when studied at the U.S. National Museum at Washington, D.C., they were discovered to contain numerous samples of sponges, brachiopods, mollusca, trilobites, and custids from the lower Silurian period, and they added greatly to the geological knowledge of Baffin Island.

From this camp some of the students after a day-long hike to the west returned to report they had discovered a ten-mile-long lake and named it Lake Porter. At the time Porter was probably quite flattered. Years later he wrote, "To the many thousand square miles of unknown territory I have since surveyed and mapped, other names have been applied. It is not considered good form for an explorer to name anything for himself." [2] However, the name was not officially accepted: the lake has no name on the map of Frobisher Bay produced by the Canadian Department of Energy, Mines and Resources (1971).

On August 26 the party started the return trip to Cape Haven, following the west side of the bay. On September 4 they reached

Watt's Bay and made camp on the northern shore. Across the bay to the southwest they saw a small crystal-blue glacier coming from a miniature icecap and hemmed in on either side by high ridges. Porter named it Boas Glacier (again the name does not appear on the Canadian Department of Energy, Mines and Resources map). He and Shaw climbed the miniature icecap (obviously Grinnell Glacier, of which Porter's Boas Glacier is simply one discharge). During their climb Porter and Shaw placed a bamboo pole on the glacier and lined it up with cairns erected at its edge. They hoped to return the following summer to measure the glacier's motion.

At the Watt's Bay camp Porter gained the reputation among the Eskimos of being a great doctor. They had discovered an abundance of ground blueberries, and one of the wives came down with severe pains suspected of being nothing more than an overdose of berries. Putting on his best professional expression Porter took her pulse and returned to his tent. Looking through the emergency kit that had been prepared for his arctic use by Dr. Cook, he came upon a bottle relating to heart and kidney problems. Because there was no better alternative, he administered a dose from the bottle. By the next night the patient had fully recovered. "I am convinced that those pills were far from what she needed—they might have killed her. But the psychological effect on the natives was incontestible." [3]

From Watt's Bay the party crossed Frobisher Bay, passing near the end of Gabriel Island, and, by way of the familiar Bear Sound and Lupton Channel, returned to Cape Haven, arriving on September 12. The *Hope* was already there with the largest of the northern Greenland meteorites on board, an iron meteorite estimated to weigh between 90 and 100 tons. But Porter was met with bad news. Disease had killed off most of the sled dogs belonging to the whale station keeper. There would be no way for Porter and Shaw to go into the interior the following spring because no others were available. There was nothing to do except call off their first chance at original arctic exploration. Porter and Shaw made the hasty decision to stay with their student companions and join Peary on the *Hope*. The entire party left early the next morning just before pack ice closed in on the shore.

Porter returned to Boston and again took up his life as a student at MIT and an employee of Despradelle.

It may seem that throughout much of Porter's college career he

was concerned only about the Arctic. But in fact his other interests
were producing results. Despradelle had designed a 1,500-foot-high
"Monument to Progress," and Porter drew a large mural of it. The
mural was displayed for years in the MIT School of Architecture
near Copley Square, and it won him an award from the Beaux Arts
Society of New York. In the following year, 1897, Porter won a
second award from the society, the first honor and gold medal in a
nationwide contest for the best architectural design for a governor's
mansion in a state capital. He was also a staff artist on *Technique
'96*, the MIT yearbook, and a member of the Boston Architectural
Club.

But Porter's thoughts never wandered far from his arctic goddess.
He was careful to get all he could out of the engineering and surveying aspects of his architectural courses, because these would prepare him for surveying and navigating in the Arctic. He once told his
friend Ralph Flanders, who later became a senator from Vermont,
that on coming home to Springfield, Vermont, during winter vacations he would get off the train in Bellows Falls and walk twelve
miles north over the hills to Springfield in order to get used to the
snow and cold for the Arctic. He used snowshoes to travel in the deep
snow and a compass to navigate through the forests.

Few facts have survived concerning Porter's years at MIT although thoughts of studies and a career were no doubt clouded by
visions of the Arctic. In the spring of 1898, however, these visions
were temporarily put aside. Al Shaw and another student from the
previous year's trip to Baffin Island decided to leave Boston for the
Alaskan coast to take part in the Klondike gold rush. Although
Porter resisted this adventure, Shaw's brother Ernest became eager to
try his luck and talked Porter into joining him to seek gold in the
rough and wild frontier of British Columbia.

The brother and another victim of wanderlust, Ambrose Attwood,
headed west to purchase supplies and ten cayuses, the small and
durable horses used by the Cayuse Indians. Porter came later by
himself on the Canadian Pacific Railroad and met the two green-
horns in the little town of Ashcroft, British Columbia. Shaw and
Attwood were camped three miles back in the hills learning how to
handle the horses and how to throw the diamond hitch. Finally,
with some mastery of the art of pack-trail life, the three young men
from the East set out north on the Telegraph Trail.

Progress was slow and life became a routine, but at least it was not the life of city routine. Rounding up the horses, breakfast of bacon and bannocks (fry-pan bread), the trail ride, dinner of beans and more bannocks, maintaining the equipment and battling ever present mosquitoes and flies—that was the routine. Eventually they passed Quesnel and continued on until they reached Lake Stuart and the Hudson's Bay Company trading post. North of Lake Stuart, many years earlier, at a place called Manson Creek, gold had been found and successfully extracted, and now mining companies were exploring more economical means of removing the metal. Porter's party continued to this location with the hope of finding their share of gold. Porter also hoped to do some mapping in the little-known territory north of Manson Creek.

While there Shaw received a letter from his father advising him to join his brother, who was having some luck on the coast. Ambrose Attwood decided to go out with Shaw, and the Boston three disbanded their expedition and divided up the remaining supplies. Porter was now free to do as he pleased, and he made no effort to hurry back east. For about a month he lived with the miners, watched them sluice the gravel in the creek beds, and did some mapping.

In such an atmosphere gold fever is highly contagious. Soon Porter had drawn up a great plan to return early the following spring on the snow and set up his own claims. But the scheme was dropped after he returned east and discovered that the financiers wanted at least seven hundred per cent profit. Sometime during September, Porter decided it was time to return home. Although experience and adventure were won from the summer, the only tangible treasure he brought back was a small vial of gold dust, which he kept as a souvenir. Years later the contents were lost when his young daughter broke the vial at play.

Porter returned to Boston and spent the winter of 1898–1899 in an uneventful way working for Despradelle. By spring the glitter of golden dreams had tarnished. Once more the arctic fever had settled into his system. He had not forgotten Baffin Island. He still dreamed of being the first white man to explore its unknown interior. But his father was showing signs of failing health, and Porter dared not go to the Arctic lest things take a turn for the worse. Instead he was asked by a Boston merchant to arrange a voyage to his Frobisher Bay whaling station, taking up supplies and bringing back bone and oil. So Porter chartered the *Lily of the North,* a two-masted schooner from St. Johns, Newfoundland, skippered by one of the famous Bartlett family.

Porter's favorite niece, Miriam Marsh, of Springfield, Vermont, who was about his own age, had been following his arctic exploits with more interest than anyone else in his family. He decided to take her on the first part of this voyage. They met the schooner at Halifax, Nova Scotia, where Porter oversaw the loading of the cargo, and accompanied the party as far as Sydney, Cape Breton Island. "The cruise in the cramped quarters of the schooner was a strenuous one for my niece but she still treasurers the experience and considers herself a survivor of a real arctic expedition." [4] Porter and his niece then returned to Springfield to watch over his father.

Frederick Porter died that summer. On the day of his father's funeral, Porter received a letter from one of his friends aboard the *Lily of the North,* reporting that she had piled up on the rocks off the coast of Labrador, her cargo a total loss. Porter never tried such a venture again. But he had not yet recovered from the arctic fever.

Of course there was the small problem of transportation. Peary was already in the North in the spring of 1900, and Porter could not communicate with him. Through the secretary of the Peary Arctic Club, H. L. Bridgeman, Porter arranged to have another expedition of students taken to Greenland on the ship *Diana,* which was carrying supplies to Peary and bringing back news of his travels. By advertising again in the school newspapers Porter rounded up six students, including his old friend Al Shaw, who were willing to pay his price of $600.00 apiece for a summer's outing in Greenland. This allowed Porter to pay off his complete college debt of over a thousand dollars and cover Peary's commission.

Thus in the summer of 1900 the *Diana* sailed north through the Labrador Sea and entered Davis Strait, which separates Greenland from Baffin Island. She stopped for a courtesy call on the Governor of Greenland at Disko and then continued up the coast and entered Melville Bay. "A quantity of enormous icebergs fill it, the product of the greatest glaciers in the world. There is also a shifting, treacherous ice-pack." [5] The *Diana* maneuvered through the bay and reached Cape York and Whale Sound, where Porter's party was to spend the summer among the Whale Sound Eskimos.

Russell Porter, center, and his favorite niece, Miriam Marsh, of Springfield, Vermont, circa 1899

Whale Sound Eskimos, Greenland, 1900
  Porter photograph

Porter water color, from photograph above, with notes taken on colors

Whale Sound Eskimos

Before the *Diana* continued north, a walrus hunt was undertaken to supply Peary with winter food for his dogs. It also supplied the students with many stories to be told back home. When they took their first walrus, "the natives went straight for his stomach," Porter wrote. The contents of clams or mussels "were considered a great delicacy, among the natives but, although some of the party tried them, I couldn't bring myself to the point of experimenting." [6]

After the walrus hunt various parties were organized to hunt caribou. Porter and Shaw went out and spent over two days stalking before they brought down one caribou. By that time they were so hungry they did not bother to build a fire but ate the meat raw. During the hunt they learned from the Eskimos that since the white explorer began coming to their shores caribou and other game had been killed off at an increasing rate. It was becoming difficult for the Eskimos to get what they needed for food.

Porter's hunting party in Greenland, 1900

Walrus hunt in Greenland, 1900, to supply winter food for Peary's dogs
Porter water color

Whale Sound Eskimos, Greenland, photographed by Porter in 1900

Toward the end of the summer, Porter went with a group of Eskimos, dogs, and sleds to explore a route that took them onto the inland icecap. Only one day's travel was needed to remove them from all traces of land, where nothing could be seen but the sky and where the snow and ice reached out for thousands of square miles across Greenland's great icecap. While Porter was away, Peary made a short and unexpected visit to the main camp. Thus Porter missed meeting him. The summer now spent, the mission of the *Diana* now completed, Porter's Greenland party boarded the ship and made its way south and eventually to Boston.

For six years Porter's arctic goddess had been beckoning him, inviting him, but discreetly never completely lifting the veil. He must continue to court, to steal the very heart of this maiden. At the time, the culmination of all his arctic aspirations was the North Pole. Porter wrote, "I must have the real thing—a try at the pole itself— the opportunity came that very winter." [7]

The winter of 1900–1901, spent in Boston, was the turning point of Porter's arctic career. A wealthy New York businessman, William Ziegler, had decided that it was only necessary to supply enough money and the Pole could be found where all other attempts had failed. He was financing such an expedition to be commanded by Evelyn Briggs Baldwin, who had been with Peary in 1893–1894 as meteorologist. Porter wasted little time in signing on as artist and surveyor to the Baldwin-Ziegler Polar Expedition of 1901.

Most of the efforts to reach the Pole had been based in Greenland and northern Canada. Few men found encouragement in the land of the eastern hemisphere, Spitsbergen and Franz Josef Land, whose northernmost point was farther south than the tip of Greenland or Ellesmere Island. Nansen's three-year drift in the *Fram* in the polar basin had not shown any land between Franz Josef Land and the Pole. Peary had said that this "eliminated the entire Siberian half of the polar basin from any further serious consideration as a possible route for reaching the Pole." [8] Nevertheless Baldwin was determined to use this route.

From the whaling fleet of Dundee, Scotland, he purchased the *Esquimau*, a 466-ton steam yacht, and rechristened her the *America*. The Ziegler finances were abundant, and no details were left unattended. With some overoptimism Baldwin purchased the auxiliary ship *Belgica* and gave orders to land a house and stores on the west coast of Greenland in case the return from the Pole was made in that direction. A third ship, the *Frithjof*, was loaded with three years' surplus food. There would be dogs and Siberian ponies complete with Russian drivers. The ship's crew would be Scandinavian. In all it was an expedition of about forty men. Baldwin had to arrange for reindeer skins, tons of pemmican, dog biscuit, dried fish, condensed human foods, and compressed baled hay.

During Porter's crossing of the Atlantic he obtained permission from the captain of the liner to use his own sextant to practice his navigation. This would be all important once the expedition was far away from the *America*. He was turned over to the third officer, David Pearson, and the two men spent many hours together on the bridge. A friendship formed that lasted for many years.

While the ship was being outfitted in Dundee and members of the party were arriving, Porter and one of the surgeons, Dr. Verner, spent a week cycling. They traveled about the eastern highlands of Scotland, visiting Aberdeen, Balmoral Castle, Spittal of Glenshee, Dunkeld, and Perth. Finally everyone was assembled, and the *America* set sail for the coast of Norway, stopping at Tromso, then rounding the northern tip of Norway and returning south to the White Sea and Archangel, Russia.

Finally, with all supplies, ponies, dogs, and men on board, the expedition sailed north. It took over a month to travel the thousand miles to the Franz Josef Archipelago. Near the islands the ice was closely packed, and it was necessary to wait for leads to open up before northern progress could be made. Sometimes the leads turned into dead ends and the *America* had to retreat south to search out a new channel.

Finally a stop was made at Cape Flora on Northbrook Island, one of the southernmost islands in the group. Wishing to spend the winter as far north as possible, Baldwin was not content to stay there and ordered the ship forced north through the ice among the islands. Actually few northward miles were attained, less than thirty, before the descending winter night made it necessary to halt on the south side of Alger Island. On shore the men erected eight-sided huts for the Russians' quarters and a stable for the ponies. The dogs were staked outside. The *America*, which was anchored a hundred yards off shore, was used as quarters for the rest of the expedition members. In honor of his sponsor, Baldwin named the camp Camp Ziegler.

Porter drawing of sled dogs on the Baldwin-Ziegler Polar Expedition,
Franz Josef Land, 1901–1902

During the long winter night the men were kept occupied by hauling back to Camp Ziegler supplies that had been dumped earlier on the opposite side of the island. Baldwin reasoned that these supplies would do double service: provide the men with training and exercise for the spring sledding and bring the needed stores to camp. Alger Island was no more than six miles along its greatest dimension and the sledding trail roughly eight miles in length, a modest route but one potentially dangerous in darkness and arctic storms. The men were not allowed to carry emergency overnight gear as this would have reduced their carrying capacity. As a result two parties became lost and suffered frostbitten feet and fingers. They were extremely lucky to be found by a rescue party. When the men protested, Baldwin refused to grant their request for emergency equipment and thus began the hard feelings between the men and their commander.

While most of the men were hauling supplies to Camp Ziegler, Baldwin took the Russians and established an advance camp, called Kane Lodge, some fifty miles north. It was to be used as a stop-over point when enough light returned in the spring to start moving supplies northward over the ice another fifty miles to Cape Auk on Rudolph Island, the northernmost of the island group. The final tally showed twenty tons of food cached on Rudolph Island. It did little good on this expedition but was useful to the reorganized expedition the following year.

One of the dogs differed from the other dogs of the pack, and Porter had decided to claim him as his own sled dog, naming him Bismark. He was the largest dog by far, with a trace of bulldog showing in his face. His hair was neither long nor curly and his coloring was a salt and pepper mixture, blending into spots of white.

Bismark, who had been recruited from Porter for these supply-hauling sled trips, never made friends with other members of the expedition or with other grown dogs. But he took great pleasure at playing with the puppies. He soon gained the reputation of being the toughest character among the dogs, a distinction demonstrated by numerous fights. During the spring sled trips he was even seen snapping at his running mate, not being able to stand a dog at his side who would not pull his share of the load. He was even accused of having caused the death of one of the dogs. In spite of Porter's efforts to laud his virtues as a strong and devoted worker, enmity slowly grew against Bismark.

Porter drawing of himself in a bear skirmish in Franz Josef Land, 1901–1902

Because even the scientists were drafted to haul supplies and thus not allowed to carry on their intended observations, Porter was surprised to find that he was left pretty much to his own activities. He was able to carry on his duty as artist and record the trail life as the light permitted. "There surely was enough happening along the freight route to keep me busy," Porter wrote. "Life on the trail, dog fights, getting around water holes, making camp and breaking it–always plenty of action. The region, too, had its charm in form and color."[9]

Sometimes, either by himself or with a campanion, he would investigate the walruses sleeping by a water hole in the channel ice. More than once he got into trouble with another species of arctic life, the polar bear.

One day, for example, sketching the shoreline of a channel, when the other men were asleep in camp some three miles away, Porter took up a position behind an ice hummock to protect himself from the wind. He was concentrating on his sketching when suddenly a bear appeared about 100 feet away. Porter made himself as small as possible behind the ice hummock, but the bear discovered him. Porter recalled that "with the customary curiosity of a bear, he ambled up for a closer look." Porter grabbed his Marlin but it refused to work. "Each time I pressed the trigger the hammer moved down slowly and barely rested on the firing pin." At thirty-five degrees below zero the oil had congealed. He dropped his gun, pulled out his knife, and began dancing and yelling, but that "had an effect on the bear the exact reverse of the one desired."[10] The bear pricked up his ears and started toward him. Porter backed off toward camp, watching the beast smell his abandoned sketch book and knife.

By this time he was again struggling with the rifle, and for the second time in his life he "tasted copper." Then, noticing that the trigger guard was not working properly, he struck it over his knee and pulled the trigger. "The bullet passed through the bear's belly and the gun kicked me in the same region."[11] Bruin wheeled around and made off down the shore, but Porter managed to throw in another cartridge and drop the animal.

During the excitement Porter's mitten had come off and the wind had done its work. Four fingers on his left hand were white as alabaster, "hard, brittle and semitranslucent." The frostbite apparently healed successfully, for he never mentioned it except in his manuscript, *Arctic Fever*. "You may be sure that thereafter I saw to it that my rifle was always in working order."[12]

The artist who goes north must be prepared to face many challenges, most of them unknown to artists in the lower latitudes. During the summer months, when the temperature is generally above freezing, oils and water colors can be used with little more discomfort than numbed fingers. But the landscape is continuously bathed in sunlight, and Porter found the glare from the snow so great that he had to be careful of snowblindness. He could not, of course, wear colored glasses while he worked. During the winter when the only light came from the moon, stars, and aurora, painting was impossible. Eager to capture the aurora, Porter once tried to work with pastels under the yellow light of a blubber lamp. The failure was not recognized until the return of the sun in the spring. Then the colors were completely unrelated to the actual auroral display.

Strangely, Baldwin did not allow the men to keep diaries. But Porter wrote years later that the expedition came to a close with little accomplished except the moving of supplies to Camp Ziegler and then north to Rudolph Island. Toward the end of the summer the men blasted the *America* out of the ice and eagerly pointed her south to Aberdare Channel and Tromso, Norway.

Not until reaching Tromso was Porter able to receive the letter containing the news of the death of his mother on Christmas Day. Before Porter left on the expedition his mother had expressed her sorrow and the conviction that she would never again see her son. "It is my one most bitter pill, my greatest regret, that my passion for the north should have brought sorrow to this dear woman."[13]

The expedition over, Porter and Dr. Verner, who had cycled with him in Scotland a year earlier, cycled south from Tromso. They traveled through Norway, then into Sweden to Stockholm and Göteborg, and finally on to Denmark. After spending a week in Copenhagen Porter went to Jutland to visit his friends from Greenland, former Governor and Mrs. Knudsen.

Then Porter again returned to Boston. He spent the winter of 1902–1903 working for Despradelle and living on West Cedar Street on Beacon Hill. Even though he considered the Baldwin expedition a complete failure, he was determined to have another try at the Pole. Within a month of his return he signed on as assistant scientist to the reorganized expedition, the Fiala-Ziegler Polar Expedition.

III. THE FIALA–ZIEGLER POLAR EXPEDITION

For Russell Porter the Fiala-Ziegler Polar Expedition began in May 1903, when he sailed from Boston to Liverpool, England. This expedition, too, was financed by the New York millionaire, William Ziegler. Anthony Fiala, in his second attempt to reach the North Pole, was the commander. William J. Peters was second in command and in charge of the scientific work to be done under the auspices of the National Geographic Society. Porter, hired as assistant scientist and artist, later became third in command.

Porter was thirty-one years old at the start of the expedition. His black hair was beginning to recede, and his blue eyes shone through signs of years spent at struggle in the Arctic. But he always gave the appearance of a gentle yet determined man who knew what he wanted from life.

From Liverpool, Porter traveled to Newcastle, then to Bergen and Trondheim, Norway, and finally to Archangel, Russia, on the White Sea. Here he bought furs and clothing, searched for his vertical transit, missing since his return from the Baldwin-Ziegler Expedition the year before, and joined the rest of the expedition on the reconditioned *America*. Among the pack dogs, Porter was glad to see his old canine friend, Bismark.

The *America* entered the sea ice under an overcast sky on the Sunday morning of July 26, 1903. Shortly after, Porter took up his pen in an attempt to explain, perhaps mostly to himself, why he was returning again to the North. He thought back over the nine

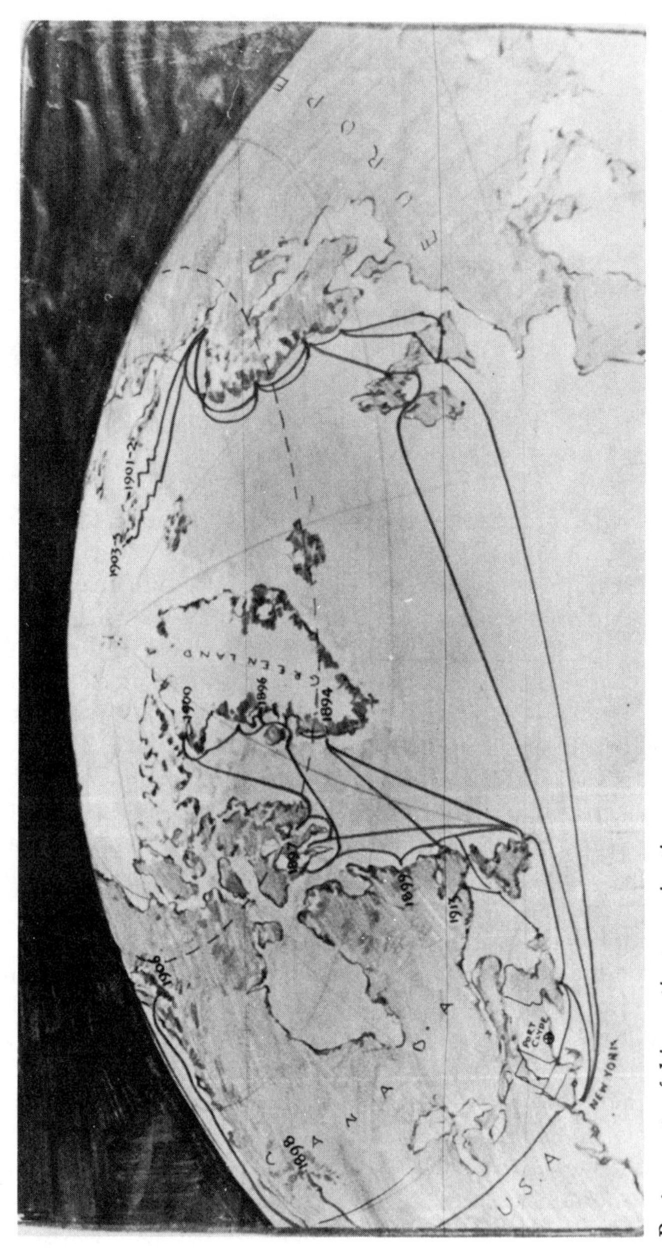

Porter map of his arctic peregrinations

years to the time when from the deck of the *Miranda*, he had first seen what he considered another world. Since then he had traveled many artic miles. But were they worth it? There had been so much disappointment—shipwreck, failures, and accidents, yet here he was, returning for another year at least. He could have stayed at home to pursue a career, and "it would be thus were it not for that germ, that something, which draws us ever back to see again, to hear again, to feel again that silence, to look again on those lands, slumbering under that 'other world' light of a midnight sun."[1]

He was convinced there was something extraordinary in the North known only to men who had made the trip and been drawn to its frozen heart, victims who could not stay away. "It is as futile to explain your affection, or a love, for this arctic maid, as it is for a human heart to explain your affection for a human soul."[2]

Northward progress on the *America* was slow. Many of the men had been with Baldwin two years before, and many arguments erupted concerning the progress then and now. Each man had his own idea of how he would force the vessel through the ice were he captain. Commander Fiala finally put a stop to the arguments by pointing out that they had entered the ice far to the south of the latitude where they had encountered it two years before.

On Wednesday, August 5, after spending several days imprisoned in the ice, unable to make any northern progress, the ship headed south, then west, then into an ice-free channel leading northwest and then north again. After this spirit-lifting advance, observations of the sun showed them to be about fifty miles from Cape Flora on Northbrook Island, which gradually rose to 1,000 feet above the sea. At six o'clock in the morning of August 7, convinced that land surely must be seen as soon as the fog lifted, Porter climbed the ice-encased rigging to the masthead. The curtain of mist slowly lifted. "All this land and ice hung over the distant pack like a vision, in deep blue and light pink, as if it would disappear should you turn the eyes away for an instant."[3] Familiar capes appeared, including Cape Flora; then dark low-lying shapes became complete island profiles.

But land was not reached on that Friday. Too much ice lay between ship and shore. More days of waiting were necessary until leads opened wide enough to allow the ship to pass. During this idle time, supplies to be left at Cape Flora were piled on deck. These

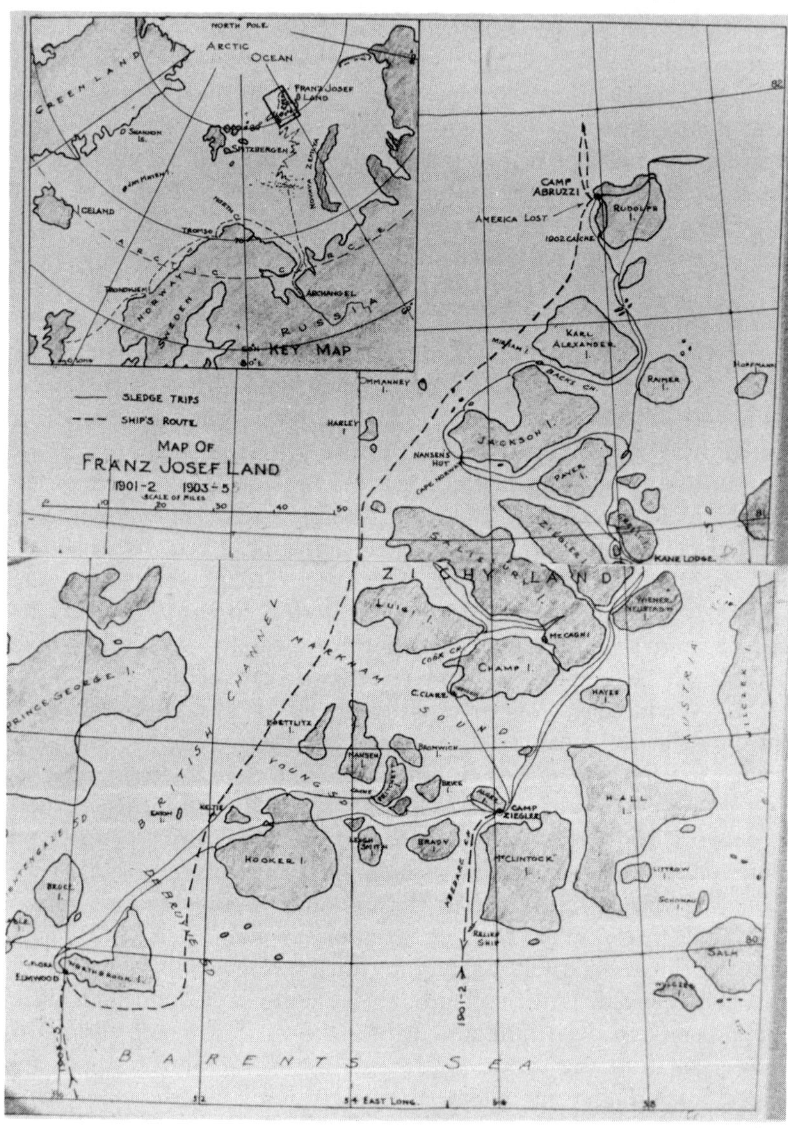

Porter map showing routes traveled and camps in the Franz Josef Land archipelago

The deck of the America *as she steamed northward through the ice-filled* Barents Sea *toward Franz Josef Land in 1903*
   Porter drawing

included medicine, ammunition, rifles, tobacco, and some food. Porter noted in his diary that the Duke of Abruzzi, passing the same point of land on his way north, had left eight months' supplies for his men in the event of any emergency. Porter added that it might be wise to continue such a practice. But little food was stored.

Among the shore party that landed on August 12 at Cape Flora, Porter found little changed since his visit two years before. Evidences of past expeditions were still visible scattered about the beach. A simple marble monument erected by the Duke of Abruzzi to honor three of his men who never left the Arctic, the remains of huts built by Jackson in 1901, a stone fireplace where Leigh Smith's shipwrecked English party had wintered over about twenty-six years before, all helped create a scene of desolation and bore witness to human struggle with nature. Note was made of the usable supplies of food and coal, and a few more supplies were brought from the ship and stored safely with those already on the island.

The ship left Cape Flora the same day it arrived because Fiala was anxious about reaching Rudolph Island before the ice became solid with the coming of winter. At first there was no ice in sight; then it seemed to be everywhere. The *America* made slow progress, steaming between islands, entering leads only to have them close, and waiting for the ice to shift so the ship could break away for another try. At least they were making better northing than two years ago. The men were thankful for that.

During this northward meandering Porter enjoyed studying the landscape and picking out the familiar points. He thought back to the Baldwin expedition and how the men had labored to sled supplies northward among the same islands through which they were now steaming. He saw Cape Brice and the rock where he had stopped to eat lunch on his sled trip west from Kane Lodge. So fascinated was he with the coast that he often fell asleep on deck, exhausted but not wanting to miss a bit of it. He also knew it was important to be familiar with every part of the landscape, for, as he noted in his diary, if the ship were lost he might have to retreat south over this very terrain. At other times Porter busied himself by working on his fur clothing. He was able to finish one complete set of fur before reaching winter quarters.

On the morning of August 31, the ship steamed past the 82nd parallel of north latitude in the open Queen Victoria Sea. This was a noteworthy event, for only four other ships had ever been farther north. That same day they reached Rudolph Island, the northernmost piece of land in all Europe and Asia. The only winter shelter that could be found for the ship was a coastal indentation optimistically called Teplitz Bay. Fiala considered leaving the ship farther south in a more sheltered harbor, but he decided not to because it would have meant separating the ship's crew and the field party. The bay offered no protection from the north and west winds and none from the pack ice. During the winter, ice would be blown into the shallow bay and frozen into a solid mass, threatening the very existence of any foreign object.

On shore the men found evidences of previous visitors. There were the remains of a large tent in which the Duke of Abruzzi and his men had been forced to winter over a few years earlier. There was a large cache of food, coal, and other valuable items left by departing expeditions. The cache of twenty tons of pemmican and small amounts of bacon, lard, and sausage, left at Cape Auk, four miles south of the *America* by Baldwin in 1902, was brought to Teplitz Bay.

In late September the men completed a house on a raised beach overlooking the bay. Adequate but cramped, the house was divided into one large living room and severel small bunkrooms just large enough for two or four bunks. A small kitchen completed the interior. Lean-tos were built onto the north and west sides of the building for windbreaks and for storage areas. (This double-walled insulation had been used by Peary in Greenland in the construction of arctic winter quarters.) The members of the field department would winter here, while the *America*'s crew would remain with the ship. After the men finished their house they assembled a canvas tent to protect the thirty ponies and 218 dogs. They ran an electric cable more than a mile from ship to house so the ship's steam-powered generator could provide light, an arrangement that could also signal trouble aboard the ship.

Away from the house, Porter built an astronomical observatory and set up his Repsold Circle for timing meridian crossings. In the opposite direction, down near the shore, far removed from any iron, William Peters built a magnetic detection hut to enable him to take measurements of the earth's magnetic field throughout the winter.

The tiny outpost, which Fiala named Camp Abruzzi, was located on the west shore of the island. Only a half hour's walk was needed to reach Lands End, the northwestern tip of the island. From there, the mighty shifting polar pack extended to the Pole, 500 miles away.

On October 15, the sun dipped below the horizon, not to be seen again until the following March. The approaching winter meant the coming of the storm season, and on October 22 a gale lasting three days battered the ship and her crew, and the electric light in the house went out. On the second day, the men could make their way to where the ship had been moored. It was gone. There was nothing to do but wait until the storm passed. On the fourth day, with visibility restored, the *America* was discovered steaming in from the north. She had been swept away by the storm yet miraculously had suffered little damage.

By November, the winter's darkness had descended on the island and its alien inhabitants. The activities inside the house were conducted under the yellow light of the electric lamp. During the night of November 11, the wind blew in from the south, forcing the ice in the bay under immense pressure. At six o'clock the next morning the living room light came on. Because it did not usually come on until seven or eight o'clock, the men immediately knew something was wrong on the ship. Porter lit a lantern and followed the electric cable out onto the bay. As he neared the ship he could see forms moving about the ice in the dim light. The crew was busy removing personal belongings. Fiala, who had felt compelled to live on the ship because of possible danger to the crew, had ordered the sleds ashore. He told Porter that the ship had been receiving ice pressure between four and seven-thirty that morning. Now the stern of the ship was pushed high above the ice, and ice blocks were heaped in huge piles about the rudder.

Fiala sent Porter back to the house with instructions for the shore party to bring ponies and sleds to the ship if the lights were turned out. Porter delivered the orders to Peters, in charge of the shore party, and immediately returned. In the short time that he had been absent, the stricken vessel had received another attack of ice pressure and was now listing to starboard. Water had entered the engine room, but a donkey pump was keeping the level from rising. Porter again returned to shore, because there was no way he could help the crew.

All was calm until November 21, when a gale blew in from the east, bringing high winds and drifting snow. The men in the house anxiously awaited further signals from the ship, and at 5:30 A.M. they saw a light coming over the bay ice. Hurrying outside, Porter discovered nearly the entire ship's crew, who had left the ship for fear it would sink. Instead of going into the house to warm up, they insisted on going into the tent. Porter reported his findings to Peters, who went out to find the ship cracked open and flooded, her ribs crushed by the immense pressures, her remains lifted high in the locked jaws of the boreal breast, the pack ice.

Later, Porter worked with the others to salvage the equipment still on board. Every piece of equipment was removed, for each piece might be a lifesaver in this world where there were no substitutes. The ship even relinquished her own materials: cabin walls were torn down for the lumber and felt insulation, and the engine room was stripped of all possible metal implements and parts. On shore some of these materials were used immediately. The tiny community had more than doubled its population and many more living accommodations had to be provided. The house was enlarged for this purpose and it took well into December to complete this task.

The America *shortly before she disappeared for the last time during a storm, Fiala-Ziegler Polar Expedition, 1903–1905*
  Porter drawing

Fiala was highly successful at maintaining harmony among his men during this difficult transition, partly because everyone was busy with the chores of establishing a new home and routine. Perhaps it was also for this reason that the real meaning and seriousness of the loss of the ship did not register immediately.

There was no contact with the outside world. No messages could be sent out, for there were no means of travel and no radio. No one knew of their dangerous position. Yet their only hope of enjoying again the pleasures of the warmer world lay in a relief ship. Could one reach them during the few weeks in the following summer when the ice would be broken enough to allow a ship to pass?

Everyone understood the real situation, but Porter had his own secret outlook. Here at last was real adventure, something to remove any possibility of boredom. "Has it not been ever so, that the surest way to success is to compel one to turn his face square to the face of the enemy, knowing that he can find scant relief if he turns tail?"[4] Porter was impressed by the skill with which Fiala handled his men, and this filled him with confidence of success in the spring sled campaign.

After the *America* had been completely dismembered and abandoned, Porter and Bismark often visited her when there was sufficient moonlight. The masts still stood but at a skewed angle, ghostly in the darkness. The bow rose high on the ice as if frozen forever at the crest of a storm-tossed wave. The jib boom gallantly pierced the blackness. Porter felt the awful desolation. On the two Ziegler expeditions he had lived for more than a year on this ship, and she was full of rich human memories. Once the cabin, now almost filled with snow, had offered a warm and cheerful shelter from the elements. It had been a place of stimulating discussions of arctic aspirations, with electric lights, an open grate fire, books lining the walls, timepieces ticking confidently, music flowing from the music box, and the aroma of hot chocolate rising from the depths below. Now those depths were a silent cavern that seemed endless in the feeble light of Porter's lantern. Part of the engine, quiet and cold, protruded from drifts of snow, and the lantern's rays made the shadows ominous. Porter found no comfort in this afterglow of habitation and he "lost little time in getting out under the light of the aurora."[5]

Throughout the winter, storms were frequent, the wind almost always present. During a violent storm in January the *America* secretly vanished. No one saw her go, no one ever knew whether she sank in the bay or was carried with the ice to some unknown grave.

At one point, Porter was much distressed. Fiala came to him with a plan that in the spring he, Porter, would return to Cape Flora with most of the men and be in command while waiting for the relief ship and Fiala's return. In the meantime, those lucky enough to be selected would make the try for the Pole. The reason behind Fiala's move is not known. Perhaps he felt Porter would be the most reliable leader in charge of the men's safety, perhaps he felt Porter was not fit for the tortuous struggle with the pack ice. Whatever the reason, Porter was unhappy and he did not hesitate to tell Fiala. He was just as fired up with dreams of the Pole as the next man. But he also told Fiala that if better men were chosen to continue north, he would accept the order to head south as something that was meant to be. Fiala understood and respected such feelings and agreed to keep him under consideration for the trail party. With this Porter had to be content, but he was not above wondering how it might look to people back home if he were to be ordered south.

Meanwhile, Porter was responsible for the astronomical duties of the expedition, and they took most of his time throughout the winter. This was an important assignment because the only way to determine the location of the camp and chart the islands during the sled trips was by celestial observations.

One of the greatest problems that confronted him was maintaining an accurate record of Greenwich time. Since this was before the days of radio, he could do it only with the chronometers. These instruments are delicate and accurate timepieces that nevertheless have rate errors. But one can keep track of these errors and account for them in position calculations if he stays in one location long enough to establish a history of the errors by timing star transits. The chronometers can then be moved carefully to a new longitude, and the process continued. In the days of radio this has become unnecessary.

In Porter's case the chronometers (only two were carried) had been on board the *America* for many weeks, subjected to the shocks and pounding of ice breaking, poor treatment for any clock. Of course it was impossible to take star transits while the ship was moving, so the rate errors could not be measured. Thus, in order to maintain Greenwich time, it was necessary to assume that the rate errors did not change during the time of travel on the ship.

After building the observatory at Camp Abruzzi, Porter was able to keep a record of the chronometer rate errors. To this end he timed several hundred star transits throughout the winter. The altitudes of the stars as they crossed the meridian were used to determine the camp's latitude. He used various means to determine the longitude. Moon culminations (the moon's crossing of the meridian) and the occultations of stars by the moon can be predicted very accurately. Porter compared these predictions with his own observations, the time differences indicating their longitude. In all he observed twenty-two moon culminations and three star occultations during the winter of 1903–1904. He took no holidays and no Sundays.

The observations were done with his Repsold Circle, an instrument designed to measure altitude and azimuth accurately to four seconds of arc. It contained a forty-power telescope and reticle eyepiece. The instrument was housed in the crude wooden observatory and rested on a pier of brick and cement anchored to bedrock.

The instrument had to be adjusted to scan the meridian so star transits could be timed. Once this was done, one of the theodolites was set up on the meridian to the north of the Repsold Circle to act as a collimator. This served as a quick check on the adjustment of the circle to make sure it did not move during observing sessions because of frosting and temperature changes by the presence of a warm body. The collimator was used during the arctic night, but during the daylight months it was replaced with a marker erected 6,640 meters south of the observatory on the meridian.

Observations were done under the most severe conditions imaginable. No heat could be tolerated in the observatory, for heat currents would refract and distort star images, causing inaccurate transit readings. So Porter would go to the observatory, open the trap doors to the stars and winds, and settle down for two hours or more of observing. The Repsold Circle was so delicate that it was impossible to use mittens or gloves to make the necessary settings. This meant using his bare hands, and sometimes the thermometer on the wall read thirty below zero and lower. (The lowest temperature recorded in *The Ziegler Polar Expedition, 1903–1905, Scientific Results*, occurred on November 10, 1903, and was forty-seven degrees below zero.) His hands became so cold he could not hold a pencil, and he had to devise another method of recording the transit times. With the help of the second assistant engineer, Anton Vedoe, he rigged up a crude

Porter using the Repsold Circle in his astronomical observatory at Camp Abruzzi, Franz Josef Land

but serviceable chronograph in his bunk room. This was connected by wires to the observatory, so it was only necessary to push a button to record the transit times. Porter's roommate, Francis Long, was recruited to operate the chronograph during observations.

Sometimes the cold forced Porter to stop work. Once when his lamp would not burn, he unscrewed the burner to find the kerosene frozen to the consistency of thick slush. Another time he could not rotate the circle without disrupting its alignment. Still another time the strands of spider web forming the crosshairs of the telescope buckled like snakes in the field of vision.

Sometimes he saw magnificent displays of aurora. "Like incandescent streamers they squirmed across the heavens, lighting up the snowscape to almost full moonlight. Another display was the radiating arch near the horizon. At times faint irridescent colors marked the displays."[6]

Preparations for the sled trip had been going on all through the winter months. Many of the men worked on fur clothing, all sewn by hand. By February Porter had finished a complete outfit of reindeer furs, parka, pants, boots, socks, and mittens for himself. He had also made several fur shirts for other men.

By the middle of February 1904, daylight was beginning to return, and the men whom Fiala had selected for the trail, including Porter, began exercising their pony and dog teams. The hundreds of pounds of food for men, dogs, and ponies were carefully weighed and packed on the sleds.

The trek to the North Pole would be long and hard and filled with danger and suffering. But Porter was filled with great anticipation and joy. The party was less than ten degrees from the Pole. It was equipped with efficient materials and priceless knowledge gained from previous explorers. The men were in good health, and the dog teams had been in training, pulling their sleds up over the glacier every day. The men liked and respected their commander. The final assault team was to consist of six men, six dog sleds, and five pony sleds. The others were to break up into three support parties and carry supplies as far north as possible on a predetermined schedule. Then one by one each was to return to land with just enough food to make it back to Camp Abruzzi. March 7 was set as the day of departure.

IV. TOWARD THE POLE

At ten o'clock in the morning of Monday, March 7, 1904, the group of twenty-six men, sixteen pony sleds, and thirteen dog sleds left Camp Abruzzi. But after they had gone only two or three miles the wind and drift reduced the visibility to near zero, and Fiala had no choice but to give the order to camp. The first duty was to the animals. The men unharnessed and blanketed and chained the ponies to the picket line. Then they unharnessed the dogs and hitched them to the steel ropes that kept the teams separated. Finally they were able to set up their eleven silk pyramid tents, which flapped all night in the wind. Cooking was done in one tent to avoid frosting in all the tents.

The next day, with clear sky and little wind, they made about ten miles on the almost level surface of the glacier. That second day's march was in a true easterly direction along a flag-marked route to a cache of provisions that Dr. Vaughn had placed earlier. While the dog teams paused for the pony teams to catch up, Porter studied the landscape. He could see far to the south islands they had passed last fall in the *America*.

From the cache the party headed north. Now the glacier sloped down so steeply that it was difficult to control the speed of the sleds. Collisions occurred and some sleds were broken. But they kept going until the northern shore of the island was reached at Cape Fligley (Fligely by Fiala). The wind and drifting had picked up during the day, and the shelter of a camp was welcome. The ponies were proving to be much slower than the dogs. The dog teams arrived in camp, the tents were set up, and the dogs were taken out of their harnesses before the pony teams arrived.

Cape Fligley was the northernmost point of known land in the eastern hemisphere. Porter found it hard to imagine a more desolate spot. To his artist's eyes looking northward the scene "was that ghastly white, broken only by slight gradations of light blue and grey over the rough surface of the pack."[1] To other eyes in the expedition the pack presented a discouraging picture with its horribly broken and twisted blocks of ice and pressure ridges. Still, Porter was optimistic that with hard work a route could be found by following the few level floes and cutting down the barriers between.

However, trouble developed before the men could even set foot on the pack ice. The following day the wind and drifting snow were so severe they were forced to remain in their tents. Temperatures ranged from twenty-five to thirty-eight degrees below zero. All but one cookstove became inoperative, a dangerous situation for twenty-six men. Five of the men were ill and unfit for further work. Fiala had no choice but to call a retreat. He would return to Camp Abruzzi and make a second start with a much smaller team. But he had to wait out the storm for two and a half days. On the morning of March 11, they headed back to Camp Abruzzi, arriving there the same day.

The failure of the trail party even before it reached the ice pack of the polar sea caused various reactions. Some of the men verbally criticized the efforts. Some were discouraged and did not volunteer for the second try. Others begged to be included a second time. Fiala wrote, "Assistant Scientist Porter made my heart glad with his enthusiastic expressions of belief in victory on our next march North."[2]

On March 25, Fiala set out again, this time with fourteen men, Porter among them, nine dog sleds, and seven pony sleds. After crossing the glacier and descending to Cape Fligley they stepped onto the twisted and distorted pack ice, and again Fiala had to call an early halt. That night Fiala asked Porter, Peters, and Vaughn, who was in charge of the dogs, to his tent to discuss their prospects. The rough ice required much chopping to let the ponies through and was both time and energy consuming. It stretched as far as the eye could see; perhaps it stretched to the Pole. The sleds were breaking up. Fiala was of the opinion that they should return to base and make another try the following year after improving the equipment. Peters preferred not to give an opinion, confining himself to the scientific aspects of the expedition. Porter was not so conservative. How could they turn back now while still in sight of land? Ziegler and the rest of the world would think they had not made an honest effort. They could not turn back now.

But return to base was the decision, and everyone carried out the orders—everyone except Porter and a companion he chose, Anton Vedoe. Porter was so set on pushing ahead that he talked Fiala into letting him and Vedoe explore the pack ice for a few days in order to gain information for the future. So now excitement and anticipation replaced bitter disappointment. Porter had high regard for Vedoe. He was experienced in the ways of the Arctic and a good worker and dog driver.

Both men were almost jubilant at the chance to be free of the crowd, free at last to explore the ways of the frozen country, free to pursue the dreams of many months. The traveling turned out to be no easy cruise.

The following day, after Fiala and his team started for Camp Abruzzi, Porter and Vedoe directed their dogs northeasterly. Their route was strewn with one huge pressure ridge after another. Each ridge could be crossed only by chopping a route for the dogs and sled through the chaotic maze of upended ice blocks. Deep, soft snow had drifted into the low spots and added to the struggle.

Porter was determined to push on. He was convinced that the pressure ridges were caused by the ice pack's having been forced south onto the island. Once away from the influence of land, the trail would become smoother and the way would be opened to the Pole. But now, while they were still forcing their way through the pressure ridges, only a mile or two of northing was all they had to show for a day's effort. At the end of each day Porter climbed the highest ridge to scan the icescape for the following day's course, "but found only the same jumbled mass clear to the horizon."[3]

The two men finally decided that these conditions were not going to change away from land and that there was absolutely no hope that a party could travel north. They swung their sled south and toward land. Fiala had given them orders to bypass Camp Abruzzi, continue south to Kane Lodge, checking the sledding conditions and caches as they went, and finally returning to Camp

Abruzzi. He and his men would use these caches on their return to Cape Flora, where the relief ship was expected.

No sooner had they returned to the island and made camp part way up the glacier than a drift storm began. Although it was not snowing, the wind picked up the snow and blew it along with such fury that all landmarks were obliterated. The only course of action was to seek shelter until the wind passed. The tent was snowproof, and the men felt snug and secure, talking about how comfortable they were and how pleased they were at not being back at camp. They were held prisoners for three days before the drift subsided.

After the storm they traveled south among the islands, where they found the ice smooth and fast. At each cache they paused to make note of its contents and to repair it if it had been damaged by bears. They swung eastward to the familiar Kane Lodge and the comforts of a stove. Starting north again, they discovered on a nearby island a seam of low-grade coal, which would later prove to be valuable. They passed Cape Norway on Jackson Island so Vedoe could inspect the remains of the hut in which his fellow countryman Nansen had spent the winter of 1895–1896.

Near the north shore of Jackson Island they discovered a tiny island not on their charts. Porter named the huge pile of rock Miriam Islet after his favorite niece. During all his years of exploring this was the only piece of land that he ever named.

By this time Porter and Vedoe had long passed the date set for their return to Camp Abruzzi. The trip had been entirely enjoyable, with good sledding conditions, agreeable companionship, and the accomplishment of useful work. But knowing that Fiala would be anxious, they hurried on.

Fiala had indeed been concerned about Porter and Vedoe's safety. He had sent two men to Cape Auk on Rudolph Island, but they found no signs of the overdue men. Fiala decided he must search for them himself, so he and another man headed south from Camp Abruzzi. After a thirteen-hour march they pitched their tent on the northernmost of the Corburg Islands, just south of Hohenloe Island (Hohenlohe by Fiala). By coincidence Porter and Vedoe were crossing the channel ice toward Hohenloe Island. Vedoe spotted the tent about a mile away. Fiala wrote in his official narrative, "It was a very happy reunion and to me one of the most pleasant experiences of the expedition."[4] The four men then returned to Camp Abruzzi on April 26 and began preparations for the 160-mile march south to Cape Flora.

Fiala assigned duties to every man. The men who had enough of arctic life were to march to Cape Flora and wait for the relief ship. Most of the dog teams and all of the pony teams would be used to haul food, clothing, sleeping bags, and other equipment south in the event the relief ship did not appear. Eleven men volunteered to remain at Camp Abruzzi to join Fiala's attempt to go north the following spring. Porter's orders were to take two men and go south among the islands to do some exploring and surveying and meet Fiala at Cape Flora. Peters would dismantle the observatories and bring the instruments and records south with a separate party.

On the morning of May 9, Porter, Anton Vedoe, and another man left Camp Abruzzi and headed south toward Cape Flora. Freed of a large party of slow-moving men interested only in returning to the relief ship, Porter was able to move swiftly and map the islands between Camp Abruzzi, Kane Lodge, Camp Ziegler, and Cape Flora. In his journal he modestly described the journey as being rather uneventful, much like any other spring arctic sled trip. But he made a valuable addition to the expedition's contributions to the knowledge of Franz Josef Land. In addition, on the north side of Hooker Island he made a detailed map of a harbor that would be valuable to future explorers desiring a safe winter quarter for their ships. His party spent the last ten days of June at this site and then continued to Cape Flora, arriving there near midnight on July 9, 1904.

The men arrived in good condition, the dogs were in good health, and they had returned with as many supplies as the sleds would carry, including fresh meat, pemmican, kerosene, ammunition, and other small items. Porter knew that if the relief ship did not reach them this summer the men would need every bit of food and equipment that could be rounded up just to survive their second winter.

The rest of the men had already reached Cape Flora and repaired the huts damaged by the winter's storms. But Porter found the camp split down the middle. On one side lived Commander Fiala and the field department in a hut they called Little Italy, so named out of gratitude for the large cache of food left by the Duke of Abruzzi. On the other lived Captain Coffin and the ship's crew in a hut they had named Elmwood. Each was independent and even the food had been divided.

*Porter water color of the hunting camp where he was in command of
five men to obtain walrus and seal for their second winter in the Arctic*

Porter was asked to camp in both settlements, but he wanted no part in this schism and set up his own tent. Here he felt he could best work up his field notes. The other two men of his field party also decided to remain detached from the rest of the community and set up camp in a small hut nearby.

Porter's living conditions were not really satisfactory. The fragile silk tent soon ripped under the strain of the arctic winds, allowing rain to soak all his gear. He elected to move into Elmwood with the ship's crew, even though the damp, close atmosphere and crowded quarters were extremely difficult for him. His reasons for shunning the field crew are unknown. He longed to be back on the trail with a good sledding companion.

The middle of July passed and no relief ship arrived, although numerous false sightings were made by eager eyes. Winds had blown continually that summer, forcing the pack ice against the land so that no navigable channels could be formed. Now doubts that any ship could reach them were beginning to circulate about the camp. As a precaution, Fiala had his men spend part of the summer rounding up food and fuel. They mined about twenty tons of coal from a vein discovered on the island about 600 feet above the sea. It was a poor grade brown coal, but nevertheless it would burn. Food left by previous expeditions, supplemented by fresh walrus, bear, and other game, would they hoped be enough to see them through the coming winter if need be.

Sometime after July 18, Porter was assigned a hunting party of four men and the cabin boy. They spent five weeks at a spot a short distance north of the camp. Much of the time the wind and rain made life uncomfortable, but to Porter it was pleasant to be away from the decaying Elmwood and with a smaller, more agreeable party. During the five weeks, the men secured fourteen walruses and a few seals. They lived in two tents and cooked over a stone fireplace, using peat, blubber, driftwood, and coal for fuel. When not hunting, they were working on harpoons, cutting up the meat, removing the hair, making long walrus lines from seal, drying the skins for soles, and caching the meat on the beach near the boats. This new supply of walrus meat would guarantee the survival of the dogs.

On August 9, while at the hunting camp, Porter received a message from Fiala:

During my absence from camp on a short sledge trip I propose taking this date to Cape Barents, you will be in charge of the affairs of the expedition and act as my representative.

Would suggest that you spend a short time each day at Elmwood until my return.[5]

Because Peters had not yet arrived from Camp Abruzzi, Fiala was putting Porter in command. The outcome was not wholly successful: three men left Elmwood for a sledding trip of a few days without letting Porter know. Even so, on September 1 Fiala told Porter that he wanted him to take command at Cape Flora while he and Peters returned to Camp Abruzzi to winter over and make preparations for another spring attempt at the Pole.

Again Porter strongly protested and asked to be allowed to return to Camp Abruzzi with Fiala, but for another reason. He told Fiala that he did not feel competent to deal with the split situation at the camp, that his command would generate much friction with older and more experienced men. Finally, Fiala agreed to let Porter go north with him.

Peters and his party had arrived at Cape Flora on August 31, and now all members of the expedition were accounted for. With the coming of September all hopes of seeing the rescue ship that year had vanished. There was nothing to do now except settle down for the long, cold winter's night. More coal was brought down to the camp, hunting parties secured fresh meat, repairs were made to the living quarters. A second winter in the hostile land would soon be upon the marooned men.

V. THERE IS ALMOST FOOD ENOUGH

During the first part of September 1904, the men who were to return with Fiala to Camp Abruzzi were kept busy repairing sleds and making other preparations while waiting for the loose pack to become bonded with young ice. "This thought of going again north and getting away from so many who have nothing to do but wait and wait and the disagreeable complications coming out of such idleness—" Porter wrote in his diary, "the thought of once more facing north is very pleasant to me, although I know the hardships will be coming to our small party thick and fast."[1]

But for days the young ice refused to form. With the details taken care of, there was not much to do but wait.

During this time of idleness, Porter became discouraged by his own arctic frustrations and failures. "I cannot, when raising my eyes to the scene outside, keep from reverting to the failure of the expeditions I have been associated with in this land of desolation": he penned in his diary, "it is well nigh discouraging and here I am entering on a third long night with far less prospects of doing anything than a year ago."[2]

In addition, certain physical problems were beginning to assault his frame. "I am still being tortured from time to time by my teeth, or is it neuralgia, for the pain shifts sometimes from one jaw to the other." The coming winter was sure to bring unpleasant side effects, "with this new complication added to my poor eyesight and deafness, to say nothing of a not too buoyant disposition."[3]

Although just waiting can be frustrating and sometimes uncomfortable, it can also provide time to appreciate the surroundings. The "striking part" of a sunset, Porter wrote, "is an effect coming out of nature's attempting to warm up this landscape of death and ice with blood crimson, an effect equally baffling to describe—brutal, unreal, mystic."[4]

On September 19, Fiala sent a five-man party to Camp Point on the north side of the island to check on ice conditions. The report came back that Fiala could cross the sound, the ice being about five inches thick. On the morning of September 27, Fiala and his last detachment, totaling seven men, four sleds, and thirty-two dogs, left Cape Flora, arriving at Cape Point in the middle of a drift storm. They found a message left by Peters stating that his party had crossed the sound on their way north only the day before. On the second day Fiala led his party out onto the sound toward Eaton Island, but found the ice too young and broken up. They traveled safely for about a mile and then came across moving ice. Porter's small metal-shod Norwegian sled went through and "in the attempt to drag it out again on to firm ice, the port stanchions collapsed, letting down that side of the sled until the crossbars rested on the top of the runner itself."[5] This clinched it. The men had to return to shore and again make camp at Camp Point. Apparently the ice conditions were so variable that Peters had felt it safe to venture onto the sound two days earlier, but now it would have been suicide.

In the next twenty days, three attempts were made to cross the sound, and three times they were forced back by shifting, unstable ice. Meanwhile, the length of daylight was rapidly decreasing, and so were the provisions. To help pass the time Porter and his tentmate Duffy built an igloo, demonstrating a lifesaving skill that Porter would later be grateful for. After the third attempt at crossing the sound ice, it was decided to enlarge the living quarters. A couple of rooms were dug out of a large mound of glacial ice and became known as The Tombs.

A certain amount of routine settled in as every man did his share of waiting. Duffy cooked breakfasts at six o'clock and Porter and another man took turns with the dinners at five in the afternoon. During the long evenings before sleep came, the men kept occupied spinning yarns about personal experiences.

One day Fiala shot a bear, which provided them with food and fuel for light. To take full advantage of this, Porter rigged up a drier by suspending an old biscuit tin over a small lamp of bear grease. After a whole day of heating, socks and mittens finally felt dry, although they were sooty from the smoky flame.

On October 17, Fiala and Porter had a long discussion. They realized their party was too large, and they knew their provisions were nearly gone. As a consequence Porter and three other men volunteered to return to Cape Flora. Porter would send back more supplies and remain at Cape Flora until spring, when he would join Fiala at Camp Abruzzi for the try at the Pole. At best this was a chancy agreement for Porter because Fiala would be making an early start—no later than March 10—and Porter would have to get an even earlier start, in February.

With Fiala and Peters both away, Porter was placed third in command of the expedition. Sergant Long and Captain Coffin remained in charge of the field and deck crews respectively.

As soon as he arrived at Cape Flora Porter showed the orders to Long and Coffin to make sure all were in agreement, especially about his getting provisions and assistance in leaving again for Camp Abruzzi. He then sent the additional provisions and a letter to Fiala saying conditions were agreeable. This he did the day of his return, and on that very night the first unpleasant incident arose. One of the men from the field camp came up to Porter and in a contemptuous boast said he did not recognize him as third in command and did not recognize Fiala either. Much disagreeable language filled the room before Porter quieted him down.

Sometime later Porter became concerned for the safety of his five dogs. There was talk of not having enough food for his animals. All too eager to start fights with the other dogs, Bismark was being threatened. Captain Coffin, the dog food having been put in his charge, refused Porter food for any more than three dogs on his spring sled trip. Soon after, one of Porter's best dogs disappeared. Later, another was attacked when a bear wandered into camp. Before the beast was brought down with rifle fire, Porter's dog was severely wounded, the hide being ripped off completely around one of his hind legs. If he did not recover by spring, there would be only three dogs in Porter's team.

These early incidents typified the situation throughout the winter at Cape Flora. Abundance of idleness, lack of proper foods, lack of incentive, lack of direction, all combined to create an atmosphere of tempers, fights, and to a certain extent defiance of orders. But Porter did all that was within his character to preserve at least some order and contentment within the community. He knew that the success of his spring enterprise depended on harmonious relations within the camp. Perhaps the very lives of the party were at stake also.

During the long winter, Porter realized that his deafness was increasing. At first he had more difficulty hearing the conversations passed about the room. Then he could hear only those fragments that were louder than normal. This disability put him at a disadvantage as commander. Had he been able to hear all that was said he might have been able to stifle antagonisms before they got out of hand. At one point it was reported to him by Long that Captain Coffin had been slurring and criticizing him, taking advantage of his deafness. Usually Coffin was a loud talker, a boaster, full of unbounded conceit and self-esteem, who would sometimes go on for hours bragging about his world travels and his whaling prowess. "I believe he chafes under the idea that there is a younger man over him representing the expedition,"[6] wrote Porter.

Shortages of food and other items were no doubt the source of some of the friction among the men. All food was issued in strict rations. Breakfast unalteringly meant oatmeal with nothing on it, black coffee, bread and butter. The mainstay of the other meals was meat brought with the expedition, supplemented with bear and walrus. "For one who has no need to stand much exposure or severe manual work (and there is comparatively little of this here) there is almost food enough."[7] Later in the winter they would be forced onto half rations.

By early December the Captain was the only other person that Porter knew who had any tobacco at all. Jimmy, the cabin boy, had offered to trade his butter ration for seven of Porter's cigarettes once a week, and this reduced Porter to barely an ounce of tobacco.

Often Porter could not sleep because of the long daytime hours indoors. He lay awake and planned his spring campaign, worked over his equipment discovering ways to shave off ounces, or mapped out his route. If that failed to bring sleep, he worked on a problem in spherical trigonometry or astronomy. The last resort was a cigarette. At certain times Porter was able to escape this uncivilized existence by tending to his stellar observations. It took a great deal of determination to carry on any kind of scientific work. The lack of proper food had reduced his incentive and ability to perform demanding tasks.

As the sun was reaching the nadir of its travels below the horizon, so also were the men approaching their nadir. Porter was no exception. On December 12, 1904, just before his thirty-third birthday, he made an entry in his field notes: "I have been getting rather morose, and ruminating over the effect of this arctic business upon my life," he wrote, "blasting all my prospects for a successful business career, aging me beyond all natural bounds and bringing nothing but the memory of a satiated desire."[8] He went on to complain that his hair was rapidly falling out, his eyesight becoming more and more defective, and his teeth fast going to pieces.

Realizing that activity would help maintain good morale, Porter kept himself busy and tried to provide the men with something to do. For example, he made a pack of cards of sheet zinc, weighing two pounds. To hold the deck he made a case of lignum vitae sled runners. But the face colors came off in flakes, so later in the year he made another deck, this time out of cardboard separators from a biscuit barrel. He used water colors from his art supplies for the face colors. This deck stood four inches high, and the dealer could hold only half the pack at one time. The pack was used almost continuously until it wore out in February, when Porter was enlisted to make a second one.

The intensity with which the crew engaged in their poker games amazed Porter. "With them [the cards] and by their means several pounds of bread have changed hands; for this is the commodity represented by the little bits of scrap tin that answer for chips." He wrote about the fascination of the game and how "some of these men have lost their week's allowance and run in debt for weeks to come." He could not understand their actions while everyone was "living on half rations, and some persons voluntarily taking their chances of going without the staff of life, little as we get of it."[9]

There were other activities to help pass the long hours. A wild commotion always erupted when a bear mistakenly wandered into camp, never to leave. For Porter there were his scientific work, his field notes, and his dogs. He enjoyed extended walks with Long of the field crew, a survivor of the Greely expedition.

Camp Abruzzi, Rudolph Island, in 1905 after the sun returned
Porter drawing

The winter solstice came, and the sun began its return trip from below the horizon. The approaching Christmas season of 1904 seemed to have little influence on the tempers of the men. Porter kept busy with his preparations for a Christmas celebration. He fashioned separate menus for each member of the field crew and special Christmas greetings for the ship's crew, including the Captain. The menus were created out of whatever bits and pieces he could find, some torn-edged paper and purple blotter paper. On each he painted a water color sketch of the tiny settlement in winter. He added a small ivory five-pointed star against a blue sky. A great deal of time was spent on these mementos, partly because he felt they were something that would be cherished years later, but mostly because such a gesture was in his character.

Porter and John Vedoe were invited to dine with the field crew. Food rationing was forgotton for the day, but the men were able to eat little more than a normal meal back home, so reduced had their stomach capacity become. (Porter ate very little for four days following.) On returning to Elmwood he distributed the Christmas greetings to each officer and member of the ship's crew. The day closed with a long poker game.

By the first of the year 1905, Porter had noted in his diary that there had been a definite improvement in the relations at Elmwood. Perhaps with the returning sun the men were anticipating the approaching season of activity and eventual rescue. Nevertheless, minor fights and arguments still broke out.

During January Porter became increasingly concerned about the physical condition of himself and the other men. They had been existing on reduced rations and inadequate vitamins for so long that their ability to work and keep warm outdoors had been drastically reduced. A walk of no more than five miles left Porter and Long weak and limp, feeling as if they had walked twenty miles. Fatigue had also lowered their resistance to the cold, and a twenty-five-below-zero temperature penetrated as though it were fifty below.

Had his condition so deteriorated that he would be unable to make his trip north to meet Fiala? If he did manage to get to Camp Abruzzi, would he be able to join Fiala on his last attempt at the Pole? He was detemined to try even if only to get away from Cape Flora.

Porter planned to leave the camp in mid-February as soon as

Porter drawing of sledding with one companion in 1905

there was enough light to allow safe travel. In the meantime he kept busy testing out a new stove, calculating how much fuel two men would need for a month, figuring out how to reduce the sled load, and watching out for his dogs.

For his companion Porter chose the fireman from the ship's crew, Duncan Butland, who had had arctic experience with Peary. Together they packed 75 pounds of food for themselves and 100 pounds of walrus meat for the dogs. This would sustain them for twelve days, enough time to get to Camp Ziegler, half way to Camp Abruzzi. There they could pick up more supplies. The sled and load weighed in at slightly over 400 pounds; each dog would have to pull over 100 pounds. During the preparations the two men had practiced until with the aid of a snow knife they could put up an igloo large enough for both of them in an hour and a quarter. This was a safety measure they would not regret.

By February 17, everything had been prepared. Porter was eager to be off, because he had to meet Fiala some 160 miles north no later than March 10. The weather was not cooperative. The next day there was little light and the wind was blowing so hard that the drift reduced visibility nearly to zero. But they were determined to start their trek on that day.

Pushing through the drift they soon left the island that had been their home for so many months and headed out onto Gunter Bay. Somewhere in the middle of the bay they threw up their first igloo, and darkness soon overtook them. There the two men and four dogs remained stormbound the next two days. Then at three o'clock in the morning of the twenty-first, they started off again in the light of a full moon, a new experience to Porter and one with which he was not completely at ease for some reason he could not explain.

At eight o'clock they reached The Tombs, where Porter had spent three weeks the previous fall. Now they spent only enough time to breakfast and then headed out onto the frozen sound. This was the red-letter day they had anticipated for so long: "Around noon time, the upper edge of the sun, like a splash of dull red molten metal, rolled along the horizon and sank again," as Porter later wrote in his diary.[10]

The next day mist made visibility on the sound so poor that they had to rely on a compass for direction. But there was little wind, which made the going comfortable even though the conditions under foot were rather rough. By nightfall, they neared Dundee Point across the sound. Too tired to build an igloo, they put up their tent instead. That night, of course, brought strong wind and drift, and the men were forced to withstand the discomforts of a flapping silk tent. By morning, travel being impossible, they were willing to accept the virtues of an igloo. There they stayed from the twenty-third to the twenty-fifth. Of the eight days from Cape Flora, four had been stormbound. This was disappointing and alarming, for there were provisions for twelve days only and the distance to Camp Ziegler was not yet half covered.

On February 25, the men continued their journey around the north end of Hooker Island to what Porter took to be Cape Markham. Here they built their next igloo on the ice foot, and none too soon, for just as they were finishing, the wind and drift came on again. After breakfast the next morning, Butland discovered that the drifting had continued all night and the snow had been piling up at an alarming rate. He shoveled out the dogs and turned them loose. "Later on we let them all inside the igloo, and they commenced to groom themselves until they were perfectly dry and apparently very comfortable."[11]

In the afternoon they tried to light the alcohol stove to make tea, but one by one the flames went out. Then Porter noticed a suspicious darkness that appeared to be creeping up the outside wall of the igloo. He said his breathing was becoming more labored, but Butland only laughed. It was obvious to Porter, however, that they were being buried, and he gave the order to break out through the roof.

On emerging from the hole in the top of the igloo they found the snow level to be just there, at the top of the igloo. After extracting the dogs they blocked the escape hatch and marked the site with an eight-foot ski sticking firmly in the snow. The sled and most of the provisions lay outside the igloo under seven feet of snow.

The men then went down among the crushed-up ice along the ice foot and put up a second igloo. Returning to the first igloo in near darkness, Butland re-entered the submerged hut and passed out to Porter what belongings had been inside, including the dog harnesses. Most of the food was still lashed to the sled. As Butland tried to break through the wall to get the sled, he heard and felt a slump in the igloo walls. Porter told him to come out before it was too late, and so the sled remained buried. Both men were now exhausted, hungry, and discouraged, and they thought it best to return to the second igloo, cook a warm meal, and get some sleep. They now realized that their choice for the campsite had been most unfortunate—on the ice foot at the bottom of a snow slope that extended to the high point on the island. The wind was blowing the snow down the slope and piling it in great drifts.

The next morning, there was little to see, and the drifting continued. Although the second igloo had been spared a burial, the first igloo was gone. Even the ski that was to mark its location had vanished. Possibly it had blown over during the night, but more likely it had been cemented in place by the drifting snow and held its height. If so, the sled was fifteen feet down and with no indication of its location.

Silently, the two men went back to the new igloo, crawled inside, put the door in place, and chinked it up. Then they each had a cup of tea. After that cheerless tea they took stock of their supplies: manfood for four days, none for the dogs; the stove, and fuel for three days; the sleeping bags, tent, and clothing bag. Most important, they had the rifle, revolver, and ammunition. They were forty miles from Camp Ziegler and about the same distance from Cape Flora.

Their lives are saved by two bears as Porter and Butland journey north
to Camp Abruzzi to meet Fiala, March 1905
Porter drawing

VI. RESCUE

Three alternatives existed. A retreat could be made to Cape Flora. Or Porter and Butland could stay and chance finding the sled before the food gave out. Or they could push on the forty miles to Camp Ziegler, relying on a food cache about half way. Each possibility had a very limited appeal. They did nothing else for the rest of the day except wait out the continuing storm.

The following morning, after a short breakfast, the decision was made: they would go on. They laid out their few remaining possessions and began to assemble them in a tight bundle, the tent on the bottom to act as a kind of toboggan. The front end was bent up and frozen in place to help in the sledding. It was just then that lady luck appeared upwind in the form of two bears, a mother and her cub. Porter and Butland acted at once. Their shots were good and put an end to the present danger, for now there was food for the dogs. The men removed the skins and entrails from the bears, gave the dogs a little meat, and cut up forty pounds to be taken along. The rest was cached among the rocks with the skins as a cover. Then the party started off again in thick fog, across Young Sound toward Camp Ziegler.

Scarcely a quarter of a mile away, they came on sharp ice blown clean of protective snow. The canvas tent bottom was torn in places, the sleeping bags exposed to the wear. Then came soft deep snow in which the dogs were unable to pull at all. One of the men went ahead to break trail while the other took up the harness to pull with

*Without the sled it took both men and dogs to pull the few remaining
supplies north to meet Fiala at Camp Abruzzi, March 1905
Porter drawing*

the dogs. Only three miles were covered that day and three miles the next. As Porter took his turn in the harness, he could sense the devotion of dog toward master: Bismark would have pulled to his death to save him. Fortunately this was not necessary, for beast saved man and man saved beast.

Hardship never really relinquishes its iron grip in the Arctic, and both men were concerned lest the sought-after cache be buried by snow. But lady luck helped a second time. As they neared the site where the cache was supposed to be, they could see an edge of an emergency box sticking out of the snow. This find assured them of replenished food and attainment of Camp Ziegler.

They were so overjoyed at finding this link with survival that they immediately made camp at the cache site. But hardship fell again. Porter was setting the last block on the top of the igloo, and the dogs had gone inside to get out of the drift. All of a sudden the whole hut caved in, Porter and the blocks piling up on the dogs. Somehow none of the dogs was hurt, although of course the hut had to be rebuilt.

By this time both men were badly frostbitten, Porter on his fingers and Butland on his face. Butland was obliged to do most of the lashing and unlashing since Porter's left hand was badly blistered from the frostbite. Luckily neither man would suffer anything more serious than blistering.

The next day they traveled with comparative ease over four miles of ice swept clean and smooth. Then it took them all afternoon to make less than a mile in the deepest and softest snow seen on the entire journey. But they finally made Bliss Island and erected another igloo. That night the last of the wooden cache box was burned to cook supper. The next morning, March 4, the last remaining candle was cut in four pieces and coaxed into providing them with some warm water into which they crumbled salty pea sausages.

Again the weather was against them, and they were forced to remain in the igloo all day because of the wind and drift. One of them thought of Porter's butter-box, which the next morning was carefully whittled to burnable pieces. Their last warm meal was prepared and eaten. The dogs were allowed to come inside, and they provided enough warmth to thaw the sleeping bags.

The following day, March 6, the drifting had ceased, and the men could see Alger Island. Camp Ziegler was only fourteen miles

Porter and Butland build an igloo on their way back to Camp Abruzzi in
the spring of 1905
Porter water color

away, and the sound ice was blown clean and in other sections packed hard with snow so the dogs could pull with fair ease. After a four-hour march they reached West Camp Ziegler and frantically tore up anything they could to build a fire to melt snow, having developed a terrible thirst from eating the salty sausage. Then the two men pushed on to the East Camp, for they knew an abundance of food had been stored there.

The camp was all but completely buried by the drifting snow when they arrived. The only things that showed were a flag pole, the top of a stove pipe, and a dark hole in the snow. That hole was the biggest surprise of the journey. They had planned to break through the roof, because they lacked the strength to dig down to the door. Now beside the hole lay a gun and shovel. Porter's first thought was that the rescue ship had arrived here last summer and left a party. On entering they were surprised to see the two familiar faces of the quartermaster and a seaman, Mackiernan, who were wintering over at this site.

After their first joyous greetings, the newcomers spent the rest of the day and night exploring the storerooms for food. They went from box to box, peering into each by the light of a candle, and brought out every kind of delicious food to sample. They got little sleep that night.

Although they certainly were glad to be in the warmth of the hut and surrounded by food, the heat soon caused Porter and Butland to suffer intensely. Their frozen fingers and toes began to thaw out, and the damaged nerve endings sent signals of pain to the brain. Butland's big toe had been frostbitten badly enough to keep him from continuing the ninety miles north to Camp Abruzzi. Mackiernan volunteered to take his place. After three days waiting out a storm and regaining his strength, Porter, on March 10, set out with Mackiernan, knowing full well that he was supposed to have been at Camp Abruzzi on that very date in order to go north with Fiala. They pushed on as quickly as possible, hoping that Fiala had been delayed.

This time they had the use of a sled, and what a pleasure it was to walk along freely with the dogs doing the pulling. There was hard snow underfoot, the sun was high enough to give warmth, and Porter thoroughly enjoyed this end of the journey. In addition, the men anticipated reaching tobacco at Kane Lodge. Porter had left a can of

Porter, right, and Mackiernan, as they arrive at Camp Abruzzi on March 17, 1905

"Seal of North Carolina" the previous spring, and now they savored the thought of the sweet aroma, the satisfying taste. On the second day from Camp Ziegler they reached Kane Lodge and went straight for the can. "Of all the impressions received above the circle, this one, the first moments with that tobacco, will remain as the most satisfying to body and mind—" Porter wrote, "a rather poor standard, you will say, to have attained after so long a sojourn there, but such it was."[1] Of course there were other comforts, like a coal fire in the stove and a hot supper with plenty of coffee.

Porter was concerned about the definite possibility of missing Fiala, so the two men left Kane Lodge soon after picking up needed food. But not until March 17 did they reach Camp Abruzzi. Fiala had left on his last sled trip north the day before, having waited for Porter as long as he dared. He knew he had not enough dogs and food to attempt a Pole dash. But he hoped at least to beat the farthest-north record set by Captain Cagni in 1900.

This was a bitter disappointment to Porter. He had struggled for so many months and then lost by so little. It was to have been his last chance to sled north toward the Pole.

On April 1 Fiala returned, thrown back a third and final time by the horribly twisted, impassable pack ice, and convinced that this was not the route to the Pole. He had reached eighty-two degrees north. This was short of the farthest-north record, but he felt he would be jeopardizing the safety of every member of his expedition if he continued on. He might be forced to sacrifice the dogs, which would be desperately needed if the relief ship failed to get through this summer, and his responsibility to the men outweighed his desire to break any record.

Now all thoughts of records and the North Pole were abandoned. Fiala's sole concern was to get his men safely south to Camp Ziegler, where the rescue ship could reach them. Porter received his orders to lead a party to the camp, pick a companion, and explore the area, called Zichey Land, that he had failed to enter the previous spring. He was then to return to Camp Ziegler and continue to make astronomical observations. Porter left Camp Abruzzi for the last time on April 17 and arrived at Camp Ziegler about ten days later.

In a few days Porter and Butland were off again on what turned out to be one of the most rewarding trips of Porter's entire arctic career. To him the party was of the most agreeable size, two. They had the chance to get away from "the cross-purposes and bickerings that go with a crowd"[2] and were granted the privilege of being the first human beings to enter the land directly north of Camp Ziegler.

At the end of the second day of traveling north with one sled and five dogs, they made camp on the summit of a low glacier. From this vantage point they were able to look down through fog into Zichey Land. Their first discovery was one of fauna and not geography: three bears decided to investigate the summit intruders, and one had to be shot to secure the camp. In fact, the land that Porter was eager to map seemed to be a natural den. Once there were six bears in sight simultaneously. During the week's trip, sixteen bears were encountered.

In spite of the local bear population Porter and Butland managed to chart nearly a thousand miles of unknown channels, bays, capes, islands, and glaciers. These were the last parts of the archipelago to be put on the map.

Returning to Camp Ziegler, Porter built an observatory out of emergency cases and materials he found about the camp and set up the Repsold Circle to take readings of the sun for time and latitude. At first he lived in a tent, but after a while he decided the observatory could be enlarged and built a bunk and table and set up a little stove. He was now living in relative comfort. In the milder temperatures of summer he was able to make better observations and in his diary noted how pleasurable it was to work under these conditions compared to the winter conditions at Camp Abruzzi. Still, the only timings he could take were on the sun because it was too bright to allow anything else to be seen.

The other men in the party were kept busy for most of the summer. Peters set up another magnetic observatory with the instruments he had brought down from Camp Abruzzi. Parties of men were sent out to obtain coal on the island. Other parties went to Cape Dillon to hunt walrus and seal, in the event of having to face another winter, and to keep watch for the relief ship.

Sometimes Peters would drop in on Porter to see how his astronomical work was coming and to look over his computations. One day near the end of July, the two men were discussing how best to use what little blank paper they had left should they have to spend another year there. Suddenly, "Chips" Tessem, the Norwegian carpenter, appeared in the doorway holding aloft a glass bottle half full of beer. Porter poured for himself and Peters, wondering all the while where the carpenter had acquired the long-absent pleasure. He tried to make a toast to the coming of the relief ship this year, but the carpenter protested, saying they did not understand.

Slowly it came to Porter. Had the ship really come? Or was this just another rumor or joke? No, it was true: the rescue ship had arrived. The hunting party had hurried back from Cape Dillon to say the *Terra Nova* had reached there. They were going to leave the Arctic after two long years. Plans were made immediately to break camp the following day.

Desperate for any news from the world beyond the snow and ice, the men eagerly questioned the hunting party as they hurriedly stored all supplies in one of the shacks and loaded the sleds for the final trek to Cape Dillon. Forewarned of much flooding on the ice, they packed the scientific instruments, records, and personal possessions in boats lashed to the sleds. The *Terra Nova* already had rescued the men who had wintered over at Cape Flora. But due to ice in the channel she could get no closer to Camp Ziegler than Cape Dillon on McClintock Island to the south.

On July 31 the men abandoned Camp Ziegler for the last time and made their way over eighteen miles of channel to the ship. It was rough going, and their eagerness to see the ship only lengthened the trip. At one point several sleds went through the rotten ice, and the dogs ended up pulling boats instead of sleds.

Soon after they started, fog settled down, enshrouding all landmarks and creating a navigational problem. But three hours out, the party met some of the men from the ship who had come to help. As evening approached a long low wail from the *Terra Nova's* siren penetrated the fog. Each man was swept up in his own emotions. Later the masts of the ship and then her massive bow emerged from the fog. Greeted with a volley of cheers and welcomes, the men wasted no time in going aboard and reintroducing themselves to the comforts and luxuries that suddenly surrounded them: letters from home, good food, wine, cigars, clean clothes, books.

From the leader of the relief expedition Porter learned that his family was well, there was no tragic news awaiting him. "The first choking sensation came on then and it has been by me more or less since," Porter wrote, "as little things would come out now and then to accent the fact of the wealth of affection still awaiting me at

The rescue ship Terra Nova, after returning the Fiala-Ziegler Polar Expedition to Tromsö, Norway, in August 1905
Porter water color

home, the efforts made to reach us in our distress, and in the passing away of Mr. Ziegler."[3]

On August 9, 1905, they first sighted Norway. On shore, one of Porter's first chores was to find a home for Bismark. He felt he could not take to the United States a dog brought up in a cold country. He found an old Norwegian fisherman living on a nearby island who promised to take good care of him. The fisherman came alongside the ship, and Bismark was lowered over the rail. The dog sat in the stern of the boat and looked up steadfastly at Porter with the troubled expression that he had seen many times in the North. Suddenly Porter could not stand that last look and turned from the rail, never to see his dog again. A year later he received word that Bismark had taken to chasing sheep and had to be shot. Even after learning this Porter paid the fisherman's taxes for several years out of gratitude for taking care of his dog for the last year of his life.

The Fiala-Ziegler expedition was Porter's final trip to the Arctic, the culmination of six polar expeditions. After it was all over he tried to justify the years spent in such enterprises. What draws a man to this white goddess, forever cold and unyielding? What takes a man away from all the creature comforts of a civilized world again and again for so many years that he sacrifices his chances to develop a normal career and to raise a family early in life? What drives a man against all natural odds at the expense of his health only to be defeated again and again? These are the questions and thoughts put down by Porter in his unpublished manuscript *Arctic Fever*. For him there were no answers. But we can say that were it not for such fevers the world might be a dull and uninteresting place in which to live.

The Porter family cabin on Breezy Hill, Springfield, Vermont, which later became the site of Stellafane, the observatory of the Springfield Telescope Makers, founded by Porter

VII. THE GREAT ONE—
MOUNT MCKINLEY

Some of the Indian tribes of Alaska called it Denali; others called it Traleyka or Doleyka.[1] The Russians named it Bulshaia Gora. Each means the same thing, the Great One or Great Mountain. Mount McKinley, as it is now called, remains one of the greatest mountains in the world. Other mountains rise higher than its 20,320-foot peak, but few can beat its actual rise above the surrounding land and its sheer massiveness. McKinley starts upward a bare 1,500 feet above the sea and soars to the highest point on the North American continent. Combine this with some of the worst weather in the world, and it presents one of the most formidable challenges to mountaineers.

In 1905 no man had yet climbed this mass of rock and glaciers that so dominates interior Alaska, and its conquest seemed to be available for the asking. Dr. Frederick A. Cook was determined to win the prize. In 1903 he had made an expedition to the mountain and had done nothing more than reconnoiter the surrounding passes and glaciers. He resolved to return in 1906 and conquer.

Needing a surveyor to accompany him in unknown country, he remembered Russell Porter from his arctic cruise aboard the ill-fated *Miranda* in 1894. Thus he was on the dock in New York to welcome home the returning members of the Fiala-Ziegler Polar Expedition. Cook took Porter to his home in Brooklyn and explained his plans to conquer Mount McKinley.

Porter was not easily persuaded to join Cook's party. He was tired from the years of arctic failure and sick at heart with the continual discouragement of the goddess of the North. In addition, his health had slowly become weakened and broken by the never-ending battering of arctic cruelties. He was aware of deteriorating teeth, frostbitten flesh, and the deafness that he had scarcely noticed in the Arctic, but that in civilization proved quite serious. Still, there was something that fascinated him in Cook's offer.

Cook went on: there would be many pack horses to carry the supplies overland. Immediately Porter did not like the sound of it, remembering his experiences in the gold fields of British Columbia in 1898. But Cook promised to have experienced packers to take care of them. Also there would be a forty-foot boat with a powerful twenty-five horsepower Lozier engine to get the men up the glacial streams. Finally, in spite of his better judgment and the thought of all the mosquitoes, Porter said he would go. It would probably be his last fling in the North.

While Cook made his preparations, Porter spent the winter of 1905–1906 with William Peters at the National Geographic Headquarters in Washington, helping prepare scientific results of the Fiala-Ziegler expedition for publication.

That winter Fiala was building up his lecture schedule, and Porter was able to talk him into including Springfield, Vermont. Thus, on February 22, Fiala presented a lecture to a sell-out audience, with Porter present as part of the exhibit. The townspeople no doubt were attracted by the wanderings of their own "boy." With his usual retiring nature, he left all of the lecturing to Fiala, who also showed slides and the first moving pictures ever taken in the Arctic. Many of the scenes showed Porter busy at some activity, and the audience loved it.

To entertain his explorer companion, Porter invited him and a few friends to spend the weekend at "The Shack," a rustic cabin on Breezy Hill west of the village. It was away from everyone and a pleasant place to relax with friends. The Porter family had owned land on top of the hill for years, and around the turn of the century Russell had helped build this refuge. Breezy Hill was to become a pinnacle in Porter's life after he returned to Springfield to live.

On May 3, 1906, Porter left family and friends in Springfield, Vermont, and headed for Seattle. The expedition as finally organized was small. Besides Dr. Cook and Porter, it included at the start Herschel C. Parker, a physics professor from Columbia University;

Belmore Browne, an artist-adventurer; and Walter Miller, a photographer, who had been with Cook in 1903. Later, twenty horses from the Yakima Indian reservation and their packers, Edward Barrill and Fred Printz, joined the party.

They boarded the steamship *Santa Ana*[2] in Seattle and on May 16 headed north. The horses were housed on the forward deck in especially made stalls. The ship made stops at Juneau, Sitka, Valdez, and Seward, but at last it unloaded the expedition and supplies at Tyonek, a tiny frontier town on the west bank of Cook Inlet. It was here that they had their first glimpse of the great Alaska Range just visible on the northern horizon. Somewhere in this vast and little known maze of peaks, glaciers, and valleys stood The Great One, 135 miles from them and still not visible.

Cook's plan was to ferry all supplies up Cook Inlet and about thirty miles up the Susitna River to Susitna Station, a small settlement of Indians and miners. But before this could be carried out, six horses that had stampeded while being put ashore at Tyonek had to be rounded up. To take advantage of the time, Porter decided to climb Mount Susitna to do some surveying.

Belmore Browne, eager to get back to the trail life, joined him. The two men were taken in the forty-foot launch, *Bolshoy*,[3] about ten miles up the Susitna River to Alexander, an abandoned Indian village. Porter wanted to travel light so they took only enough food for two days, expecting the *Bolshoy* to return in that time. From Alexander they hiked overland about ten miles and made camp at the base of Mount Susitna.

The next day the mountain was shrouded in dense mist, surveying was impossible, and there was no choice but to return to meet the launch. The appointed day came, but no launch appeared. They spent part of the next day searching the deserted village for food. There were fish in the river, but the men were cut off from this source of food by a backwater and had to settle for dead fish trapped near shore. They found that the sun-dried side of the fish was edible. Then they found some oats in one of the Indian cabins, and after roasting them they were able to brew some brownish water that reminded them of coffee.

After four days, instead of the planned two, the *Bolshoy* came plowing up the river full of supplies, men, and food. Porter and Browne headed straight for the food lockers, pausing only to meet the two new volunteer members of the party, Captain Armstrong and Russell Ball, both eager to do some prospecting in the head-waters of the Yentna River.

The delay of the *Bolshoy* was due to the horse problem. Cook had taken the launch south along the shore in search of the animals. It had run aground on the tidal flats of the inlet and they had been forced to wait out the low tide. Then they had run into storms and been forced to wait them out. They finally had to give up the search and return to Tyonek, spending another day there loading the sup-plies. Then, on June 3, the packers had started the long overland journey to the headwaters of the Yentna River, while the rest of the party had returned to Alexander to pick up Porter and Browne.

Cook's plan was to take the launch as far up the Yentna River as possible and then, while waiting for the horses, to explore the rest of the valley on foot. He hoped this valley would provide a route for the horses into the Alaska Range and to the southwest arête of Mount McKinley. From observations on Mount Yentna he had reason to believe that this might be the case.

The launch was soon on its way, loaded heavily with supplies and men. During the entire trip Porter charted the Yentna's winding course from an observation post atop the kitchen box on the stern deck. Occasionally they passed a small group of prospectors. Finally the *Bolshoy* could go no farther, and there on the East Fork they made camp. While the others busied themselves making camp liv-able, Porter measured a base line and took sightings on mounts Kliskon and Yenlo.

Then Cook, Porter, and Browne started to explore the upper val-ley of the Yentna for the hoped-for route to The Great One. They shouldered their packs and started off with only three days' food sup-ply—pemmican, erbswurst, tea, beans, and bacon.[4] Cook and Porter carried rucksacks that were too small and made their loads ride too low and too far to the rear. It was necessary for them to lean far for-ward to maintain their balance. (Modern packframes are designed to carry most of the weight close to the body and high up on the shoulders so that the packer may stand more nearly erect and expend less energy while walking.) Browne eased some of the load from his shoulders to his neck and head by using a "Russian Aleute" strap, a band that attached to the pack and passed over the forehead. This technique also made it easier to get out of the pack in case of a slip during one of the icy stream crossings.

The three men passed through a wide flat glacial valley and continued as the valley grew narrower and the sides steeper. At times they were forced to travel high on the valley walls to avoid the swift glacier-fed river. Soon they had a brief glimpse of Mount McKinley, that great white massif, master of the Alaska Range.

Their camps, as described by Belmore Browne, "scarcely deserved the dignity of the name: a small tent; a wisp of smoke from a brush fire surrounded by steaming, ragged clothes; some black pots, and three sun- and smoke-browned men hugging the fire—that was all." In the true perspective of the surrounding vastness that dwarfs all man's efforts, "we were nothing but an indistinct blur in the shadow of the mountains." As a result of the strenuous exertion and stream crossings, "We were wet to the skin constantly, and dried our clothes at night, sitting more or less naked about the fire during the process." To enhance the picture of trail life, "The wilderness too had set its brand on us; we were as dark as Indians except where mosquito bites had blotched our faces with red." By this time, "We were a rough-looking crew."[5]

At last they came to the canyon of the Yentna River. If a pass through the Alaska Range for the horses existed, it would begin here. Perhaps the success of the entire expedition depended upon that pass. The men had already been out three days and their food was running low. Cook had hoped to see the pass from some mountain top within three days, but they had been forced by the rugged snow and ice-covered mountains to stay in the valley, taking longer than he had planned. Now they still had to push forward on reduced rations, planning to find game along the way.

They climbed a ridge to the west and followed a course that paralleled the Yentna Canyon. Climbing higher and higher, struggling with heavy packs and empty stomachs, they continued until well above timberline. They had no fuel, and their food consisted of dried fruit and hardtack washed down with ice-cold water. At each camp Porter, responding to his duties as surveyor, continued to take sightings to surrounding peaks.

After a while their view of the canyon was blocked by mountains, and lack of nourishment forced them to retreat to the canyon floor.

All that remained of their food was a half-pound of erbswurst, a quarter pound of bacon, and a handful of tea, and they were several days from base camp. The men, too weak and exhausted to continue the exploration, turned for camp not knowing if a way existed through the canyon. After surviving for eight days on three days' food, the three scouts stumbled into base camp, hungry, exhausted, and much discouraged.

Still waiting for the horses to arrive, the men cut a trail through the forest, out of the valley and directly to the timberline, to help bring in the pack train. Four days were spent hacking through thick underbrush, following an old moose trail, heading for less brush-covered areas, or circling around a bird and her nest; and all the while torturous hoards of mosquitoes hung around them. Even above the timberline the mosquitoes attacked in swarms.

Porter, camping with Belmore Browne, continued the plane table surveying work and here had his first experience with an Alaskan brown bear. One morning the two men awoke and saw a bear and cub far below them in a natural amphitheater. Browne took off down the mountainside to investigate and the bear either heard or smelled him. She and her cub climbed the mountain to the tiny camp, but they did not stay around long enough to study the human intruder named Porter.

After breakfast Porter had climbed to the top of a nearby mountain to finish some survey sightings when the bear suddenly reappeared. She charged Porter, and he immediately turned and ran down the steep mountainside, past the camp, until he discovered he was no longer being chased. Scratched and bruised he struggled back up, much to Browne's surprise, because he had been unaware of Porter's descent. Browne, going up to the ridge, discovered from the tracks in the snow that the bear had soon given up on Porter and continued on to the upper snow fields.

The next day the two men returned to base camp, and the following day the pack train arrived. Only eleven horses had survived the grueling overland trip. After a short rest, Cook set the entire party on the trail to explore the Yentna Canyon and attempt to reach Mount McKinley by crossing the Alaska Range to the west.

This undertaking was not easy. The small group of already tired men and horses faced unbelievable struggles with the enraged elements. The mad glacial water of the creek was forever ready to sweep man and beast downstream, to drag them under or lash them against a boulder. But they all made it safely back to camp. Cook was forced to concede that there might not be a route through the Alaska Range for his pack train, and he decided for his next attempt to keep to the southeast slopes.

Accordingly they crossed the East Fork of the Yentna, climbed Mount Kliskon, a low 3,000-foot mountain, and were rewarded by a sweeping view of the vast Alaska Range. It was one continuous vista of towering ridges and peaks, glaciers and glistening snow fields. All eyes were captured by the master, Mount McKinley, citadel of rock, ice, and snow. McKinley, fortress armed with deadly avalanches, protected by violent and sudden storms, secure in its rarified atmosphere, how could man do battle with this giant? But the days and weeks had been passing, and there was no time for deliberation or the battle would be lost.

Now they had to drop down and travel the low country. There, swamps and marshy valleys made strenuous work, expecially for the horses. Many times an animal sank deep into the bogs and had to be unloaded and laboriously extracted with the help of the other horses. They became visibly weakened by this ordeal, and the men realized the horses could not travel much longer.

But the explorers continued on, passing a mining camp called Sunflower and then climbing again to above timberline. Finally they reached the terminus of the Tokositna Glacier. Here was an expanse of highway heading to Mount McKinley. At least this is what it looked like from a distance, but the men knew better. No place for horses, these glaciers, strewn with huge blocks of ice and scattered boulders and crisscrossed with deep crevasses, some hidden by snow bridges ready to go down under any weight put on the wrong spot.

Before starting onto the glacier, Cook, Parker, and Browne climbed a peak to the west to study the proposed route. They could see all the way to Mount McKinley, but the view was complete disappointment. Their goal was just too far away and separated from them by a glacial route that only a strong party could traverse before the end of summer. Not only was Cook's party weary; The Great One looked completely unclimbable from the south. To add to their discouragement, the nights were beginning to show signs of frost, hinting of the approaching winter.

Cook did the only thing he could. He abandoned this route and

Porter water color of Mount McKinley

decided to explore farther north and east for another, more feasible passage to his prize. They crossed the two-mile-wide Tokositna Glacier, climbed 2,000 feet to a mountain ridge, and climbed back down to another glacier. Porter used the vantage point of the high-ridge crossing to do more surveying.

The glacier on the north side of the ridge appeared to offer easier traveling, and it did flow from the McKinley direction. Because Cook wanted to use it as a reconnaissance route, the men started down under heavy packs. They soon discovered that just to get onto the glacier they would either have to cross an icy torrential stream or scale a vertical wall of ice. They chose to find a route through the latter. But not a crack could they find, even after miles of searching.

Enthusiasm and optimism for the whole venture had been slowly waning for some time, and about thirty grueling miles still separated them from their goal. When it finally became evident that they could not climb Mount McKinley this summer they decided to return to the base camp on the Yentna River and then to Tyonek.

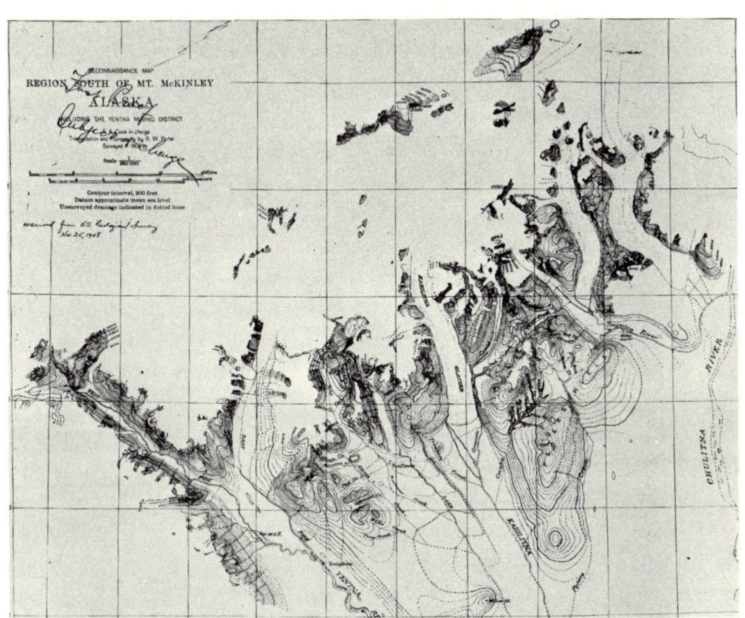

Map of the Alaska Range with Mount McKinley at the very top Triangulation and topography done by Porter in 1906

After further surveying, hunting, and reconnaissance, they headed for home and planned to return the following year, better equipped with experience and knowledge of the area.

Cook split up the party and sent Porter with two other men[6] to finish his surveying work to the west and south of Mount McKinley, including some areas never explored by man. Porter wrote, "I was several weeks clearing up three thousand miles of hitherto unknown peaks and glaciers."[7] His discoveries were later included in a separate map quadrangle published by the U.S. Geological Survey.

During Porter's several weeks of mapping, Cook took his packer Barrill and a new hired man and headed up the Susitna River in the *Bolshoy* to explore possible routes to the northeast slopes of Mount McKinley. Shortly after Porter returned to Susitna Station, Cook arrived, claiming that he had indeed succeeded in reaching the very summit of Mount McKinley, that the prize was actually his.

There must have been more than a little skepticism at camp. Could two men (the third man did not travel all the way with Cook) in the short time they had, reach the mountain, pack enough supplies to climb it alone, and return to tell about it when the entire expedition had not been able to reach even the base of the mountain in more time? Cook was a convincing man and insisted on his story. He even returned home and published a book describing his daring feat, although it did not appear until after he had left for the Arctic to claim the North Pole. Porter helped to illustrate the book from Cook's oral descriptions.

The story is Cook's not Porter's and is well documented elsewhere. Four years later Browne and Parker returned to Alaska and collected photographic evidence that they said proved without a doubt Cook had never reached Mount McKinley. Finally Barrill publicly confessed that this was true.[8]

Still, Cook was an excellent leader, and Porter was not sorry to have been with him on the Mount McKinley expedition—although it took considerable time and effort to obtain his pay. There is no doubt about Porter's respect for Cook's ability as a member and leader of the difficult expedition. "He was always in good spirits and ready for more than his share of the drudgery, resourceful and considerate of others."[9] But Porter had to admit to his friend Senator Ralph Flanders of Vermont that he had discovered his idol had feet of clay.

VIII. PORT CLYDE

By the time Russell Porter returned from the Mount McKinley trip, he was ready to settle down. Thirty-four years old, with nine arctic or subarctic trips to his credit, he was ready to give up his affair with his arctic goddess and start a new life. But once an explorer, always an explorer. Although he thought he was through searching in a geographial way, he was to continue to be an explorer for the rest of his life. Freed from wanderlust, he would focus his attention on art and optical science.

First Porter needed a place to live and a means of earning a living. He showed little interest in architecture, and he had no ambition to make money. Instead he had in the back of his mind the idea of starting an art colony.

The ideal spot for young artists to come and study and work in natural, unhurried, and inspiring surroundings would be away from the cities. Yet Porter did not want to be too far removed from their cultural and other advantages. Boston was the city Porter knew best so he chose to remain nearby. In addition, he also had acquired a strong attachment to the sea during his years traveling to and in the Arctic. The ideal compromise seemed to be "down East." After studying topographical maps and writing to real estate men in Maine, Porter chose the place where he wanted to settle down and establish his colony.

Half way down the coast from Portland to Bangor lies the town of Thomaston. Not far from here a road turns south from Route 1 and winds toward the end of a long peninsula. Passing through open fields, turning to face the sea breezes, and quietly threading its way through the little towns of St. George, Tennants Harbor, and Martinsville, it comes to rest at Port Clyde on the very tip of the jut of land. Porter would settle in Port Clyde. Having decided, he traveled from Boston to visit the town, by train to Rockland and from there south by buggy to Port Clyde. During the fall of 1906 he returned to buy a house and about fifty acres of land from a fisherman named Alfred H. Marshall. Marshall was also the town clerk, and his daughter Alice Belle, an attractive young woman, acted as postmistress.

Porter never forgot the first time he walked down the center of the tiny fishing village and into the post office. Alice Belle was busy selling stamps at the window. To the man who had spent many lonely years in the Arctic she was the beautiful young woman he had longed for all the while. She was the perfect antithesis of his arctic goddess, "alive and warm."

"Good morning," he started, "I would like to buy two two-cent stamps for these letters to Boston, please."

Alice replied, "Certainly, that will be four cents in all." After a slight hesitation she added, "You must be new in town."

Because of his nervousness and his deafness, he had to ask her to repeat what she had said. And so a courtship began that slowly developed all the while Porter worked on the house he bought from Alice's father.

The farmhouse had been built by the Marshall family in the early 1820s and had not been lived in for some years. It sat on a high point of land and offered a view of the Port Clyde harbor to the west and the open sea to the east. Only the Marshall's Point lighthouse was farther south on the peninsula. Breathing deeply of the spruce-scented air of the nearby woods, Porter thought, "This beautiful place is mine. I still can't believe it."

The art colony never materialized, but Porter kept busy repairing and improving the house. It was in rather poor condition when he bought it, had even been used to store hay from the field. On the south side of the house was the living room with its massive brick fireplace. Numerous doors led from this room to the kitchen, bedroom, den, and sitting room. There was no second story except for an attic. The cellar with the flagstone floor was almost filled by the brick foundation for the three fireplaces on one chimney. In his spare time, at the entrance to his land, Porter erected two fieldstone pillars with a bronze plaque proclaiming the site "Land's End."

During that first year, Alice became deathly ill. By now Porter was deeply in love with her and acutely anxious about her condition. After each day's work on the house, he rowed across the bay to the Marshall house on the other point to do anything he could to help the stricken family: doctor-fetching, nursing, even monetary aid. He sat long hours with Alice. Gradually the fever subsided and the battle was over. Alice won back her life, and Russell won her hand.

In the fall, plans were made for their wedding. Porter needed money to support his new bride, and on November 16 he sold his interest in his father's house in Springfield, Vermont, to his older sister, Annie Marsh, for $950.00.

Then on Wednesday evening, November 27, 1907, the wedding ceremony was held at the home of the bride's parents. At eight o'clock Alice Belle's cousin began Lohengrin's wedding march. The couple entered the parlor and stood under an arch of evergreen and red berries sparkling with fairy lights. Ever mindful of his recent past, Porter arranged that they should stand on a polar bear skin rug from the Fiala-Ziegler Polar Expedition. Only members of the families and a few guests were present, including Porter's brother Frank, of Springfield, Massachusetts.

The happy couple left that night in Porter's horse and buggy, ostensibly for Rockland, but after traveling just a short distance, they quietly slipped back to Land's End to spend the first of their honeymoon in their own house. Then they took a trip to Massachusetts and Vermont, and sometime after the first of January, returned to Port Clyde.

Soon after Porter settled at Land's End the necessity of earning a living became a reality. Because the art colony had not materialized, he decided to try farming. Other people in the area were growing potatoes for a profit, and this might be the answer. But there was always something else to steal the attention of his active and creative mind—a scene to be painted or an object to be sketched. Porter was not the agrarian type.

It is difficult to date Porter's various projects. Some were accomplished after the short-lived farming attempt, and some were done during its heyday. But exact dates are not necessary to get an idea

Porter acting the part of a squire, about 1909

of why the poor farm suffered. He built a long pier with ramp and float reaching out into the bay. He used oak pilings and cribs of spruce filled with rock. He found clay deposits on the land and built a tennis court near the house. Across the road from the house grew a large spruce tree, the tallest around by far, strange and intriguing, with large burls protruding from most of the limbs. Porter built a circular staircase about its trunk and erected a platform near the top, just to satisfy his urge to climb high and survey the land and sea. He named it the "Crow's Nest."

At last, Russell and Alice Belle had to recognize that farming was not Porter's calling, nor was there a living in it on their scale. He had been trained in architecture, but so far the arctic wanderlust had stolen any possible chance to start a career. Now Porter decided to build cottages along the shore of Land's End and rent them. If this did not utilize his advanced architectural knowledge, it would at least satisfy his creative hands and provide a source of income. He set about the task of designing the cottages and, with the help of a second man, began construction, which spread over several years. The cottages were built in a rustic English style with dormers and gables and covered with cedar shingles. The rooms were generally small, and the main room could be easily heated by a large stone fireplace and cheerful fire that reflected its light from the gray-stained woodwork. No two cottages were alike. One had a balcony that suggested a Swiss chalet. Another larger one had many gables reminiscent of Hawthorne. In all there were fourteen, each having a name. For "The Penguin" Porter carved a wooden penguin and mounted it on the outside on the stone chimney. Eventually Porter tired of playing the part of landlord and sold all the cottages.

The year 1909 saw his and Alice's greatest tragedy. Their son Marshall was born on June 13 of that year. As in any family, the first child must have brought great joy, hope, and dreams for his future. But these were not to be. Marshall lived only a few days. It was a blow to both of them, and Alice suffered the more for it, at least outwardly. In order to take her mind off their loss, Porter decided to take his wife on a European tour during the winter of 1909–1910. There was not enough money for such a trip, but to help finance it he arranged to do commissioned paintings.

Then, in 1910, Porter embarked on another journey, one that was to continue the rest of his life and develop into what was per-

Two of the fourteen cottages built by Porter at Land's End, Port Clyde, Maine

Porter water color done after European and North African tour in 1910

The Jungfrau from Wengen, Switzerland
Water color by Porter, after 1910

haps his greatest accomplishment. While spending the years in the Arctic, he had become fascinated by the art of celestial navigation. This fascination with the power of the stars was now slowly developing into a deeper awareness of cosmic beauty and its mathematical precision. How much more a person could learn of the stars with a powerful telescope! His growing interest in astronomy was matched by interest in the optical instruments used in its pursuit.

James Hartness, a good friend from Springfield, Vermont, and president of the Jones and Lamson Machine Company, was aware of Porter's interests. He himself was an amateur astronomer and did what he could to spread the contagion. One day Porter received a sample of *Popular Astronomy* magazines from Hartness, in which, quite by chance, there was a short article on how to make one's own telescope. Hartness had baited the hook that was to hold Porter for the rest of his life. But that is getting ahead of the story.

After the European tour the Porters resumed their life at Land's End. For Russell this meant a return to his varied and never-ending projects. But not all of his time was spent on the house or on building and renting cottages or doing surveying to earn a living. The wanderlust is a difficult urge to suppress. His old friend Henry M. Bryant, at one time president of the Philadelphia Geographical Society, had been to the Arctic on some of Peary's relief expeditions. Once again Porter found himself making plans for a northern trip. Bryant's hope was to go from the mouth of the St. Augustine River, in the northeast corner of Quebec, north to Hamilton Inlet in Labrador, an air distance of some 200 miles but much longer by the devious river route.

With the services of two Newfoundland canoemen and two Indian guides hired at the Hudson's Bay Company post at the mouth of the St. Augustine River, the small expedition started north on July 12, 1912, with 500 pounds of supplies in three eighteen-foot canoes. At the first portage the Indian guides deserted. Later, rapids and a waterfall made necessary a portage that an Indian trail map showed to be ten miles long. One canoeman was incapacitated by an injured knee and it took ten days to make the ten miles. Farther up river, at a fork, Porter stayed with the injured man while Bryant and his canoer tried the main stream. They found the lake and saw the height of land they were looking for, and Bryant felt that ten more days would bring them to Hamilton Inlet. But he knew the

Algiers street scene by Porter, 1912—water color

Water color of Capri by Porter, after 1910

injured canoer was suffering greatly and might receive irreparable damage to his knee if he continued. Bryant turned his party around, descended the St. Augustine River, and arrived at the Hudson's Bay Company post on August 23.

By the time Porter reached home, plans were again being made for the arrival of a new family member. On November 29, 1912, Caroline was born. It was not an easy birth; both Caroline and her mother nearly died. Caroline was the last birth in the Porter family.

There were new projects to be worked on. At some point Porter designed the town library for Port Clyde and the Episcopal Chapel down the coast in Belfast, Maine, each unique, but in the Porter style, with a wooden framework sided with cedar shingles.

While Ta, as Caroline was called by her parents, was a young girl, Porter decided that the guests who frequently came to visit required an addition to Land's End. He still had many friends in Boston who enjoyed coming to Maine. For them he journeyed up the peninsula to Rockland to buy an old fishing vessel, the *Sara A. Blaisdell*. She had been neglected for many years and, as Porter told Caroline years later, was so unseaworthy he had to float barrels under her deck for fear she would sink while being towed to Land's End. He hired some men to help him drag the vessel onto land. After a short time the workmen refused to go on, saying the boat was too unsafe to be turned into a guest house. Porter was forced to change his plans, but saved the rafters and other woodwork that were still sound. The unused parts of the hulk sank near the pier to rot.

Land's End had many crisscrossing stone walls, which hoarded abundant free building materials. He would build a stone guest house; better yet, it would be a castle complete with tower.

With another man to help, Porter began hauling stones and erecting the stone walls. Several sleeping rooms were built of wood, but the main dining and living room and a small circular room called "The Keep" were constructed of stone and lined with gray paneling. A long, low, arched window with diamond-paned leaded glass looked out over the bay. From a side door a circular staircase inside the square tower led to a view of the land from above. Another doorway led down to The Keep, which was completed with a large stone fireplace, brick floor, and brick seats with straw cushions lining the wall. To commemorate the *America*, which had been lost on the Fiala-Ziegler Polar Expedition, Porter depicted her in a stained-glass win-

Episcopal Church in Belfast, Maine, designed by Porter

The guest house built by Porter in Port Clyde, Maine

Water color of Porter's guest house, the Castle

dow, frozen in the ice for the last time, proudly lifting her bow high in defiance of the crushing ice. The roof of the castle was flat and the exposed ceiling rafters came from the *Sara A. Blaisdell*. With clay from the land the castle was completed with gargoyles in the corners of the roof.

In addition to his building projects, Porter had other things to do. Sometimes he would take the family on day trips to visit friends or some point of interest on the peninsula. His paint and pastels box always went along. His brush might capture a gull skimming the bay in search of its next meal or record a still life of bowls and freshly caught crabs on a table. Land's End as seen from across the bay, a fishing shack on Monhegan Island, the pergola porch, the fieldstone castle are typical of the scenes that became water-color impressions. Porter was equally skilled with a pencil, and he as easily dashed off sketches of the interior of one of his cottages or any other object that caught his fancy.

In addition to his art, he had his astronomy. Let us return to the bundle of *Popular Astronomy* magazines sent by James Hartness. The title of one article was "Speculum Making." The paraboloidal mirror called the speculum was the very heart of the reflecting type of tele- scope, the type best suited to amateur construction. The author was Leo Holcomb, of Decatur, Illinois. The actual method of making this speculum was not described, but Holcomb wrote that he had built and was using a ten-and-one-quarter-inch diameter reflecting tele- scope and that the hobby was fairly widespread in England. The article no doubt inspired Porter to think, "If these amateurs can make their own telescopes so can I."

Porter wrote to Holcomb and was sent the book *Glass Working* (London, Cassella) by Paul Hasluck. He then sent for the necessary glass discs and other materials and set about making his first specu- lum. Into the flagstone floor of his cellar he sank a wooden pedestal about which he could maneuver to work his glass discs with abrasives and water until they took on the required curved shape. There is some dispute and no proof as to the size of the first mirror he made, conjecture ranging all the way from a two-inch- to a ten-inch- diameter disc. But by the time the Porters were living in Springfield, Vermont, Porter had completed a dozen or more telescope mirrors, the largest sixteen inches in diameter.

Most of the mirrors never got into telescopes. There is something so intriguing about making a mirror that the maker has to start on the next one before the last is placed in a telescope. But in order to explore the celestial sky Porter had to build more than just telescope mirrors; he had to build complete telescopes and observatories to house them.

Thus in 1911 when he acquired a six-inch refracting telescope, he proceeded to build a small observatory to house it next to his home. It was built on a stone foundation with wooden walls. The dome was a wooden framework covered with canvas and weighed about 200 pounds. Because Porter never wrote how he happened to receive this instrument, one that was far too expensive for most amateurs, we might speculate that it came from his well-to-do friend James Hartness.

The little observatory came to be used less and less on winter nights, not surprisingly. After all, Porter had not forgotten the many hours spent in the arctic cold with a transit telescope. But this memory set him to dreaming about the possibility of building an

Porter's first observatory, which housed his 6-inch refracting telescope, Port Clyde, Maine

observatory that would allow the observer to remain inside, warm in winter and away from mosquitoes in summer, while the optical elements remained out in the cold, safe from distortion by heat. This dream did not materialize for some time, but in the interim the dreamer again received encouragement from James Hartness.

Hartness realized that Porter possessed the unusual knack of being able to shape glass to a paraboloid within the high degree of accuracy required to produce superb astronomical telescopes. Here was a man embodying great curiosity, dexterous optical ability, creative substance, and a skilled hand with the arts. Here was a man worth encouraging, and Hartness did much to help him along.

Once Porter received a letter from Hartness and years later wrote that it contained "a veritable wad of two hundred dollars." He continued: "Accompanying it was a card stating that he realized that I was having rather hard sledding with my work, and wouldn't I please accept this money and buy a lathe, drill press, etc., to aid me in my work." Porter never forgot this act of generosity. "I carried that little card around with me in my pocket book for years until it just wore out and went to pieces—not to show to others but just to have it handy to look at."[1]

Porter had long dreamed of building a big telescope. Then in December 1913 Hartness had two sixteen-inch glass discs sent to him from the shop of John A. Brashear, the famous American telescope builder. Another time Hartness gave Porter his old Stevens Duryea touring car, which was ready to be scrapped. It became an excellent source for mechanical parts, telescope bearings, and the like. Porter had to go to Boston to get it, and the old car did not make it all the way to Port Clyde without being towed. Once at Land's End it was disassembled and acted as a stock room for telescope parts, the envy of any amateur eager to use existing parts to save money.

Now Porter set about designing and constructing his sixteen-inch telescope. Again Hartness was a help, for the two men corresponded frequently about the designs. Another astronomical instrument that fascinated Porter was the sundial, and the two men also discussed its design. As Porter drew up plans for his schemes he would send them off to Springfield for inspection and suggestions. In this way he greatly increased his knowedge of optics and telescope design principles.

There were to be two mirrors in this new telescope, a flat mirror to collect the light and a concave paraboloid to focus it into the eye-piece. The first part of the task was the grinding and polishing of these mirrors. He planned to make the focal length of the paraboloid about sixteen and one-half feet to give the telescope sufficient power. This meant he had to have twice that length in which to set up the simple testing apparatus of acetylene lamp, pinhole, and kitchen knife used in determining the shape of the mirror as it was being polished. Working again in the cellar he found that the stone pier supporting the three fireplaces in the rooms above obstructed the light path he needed. So he cut out a notch and gained his testing space. The better part of a year passed before the optics of this new telescope were completed to satisfaction.[2]

The telescope incorporated a new design to provide for that warm observing room. At the southwest corner of the house Porter

The Porter home and observatories at Land's End, Port Clyde, Maine

The Porter home at Land's End, Port Clyde, Maine
Part of his 16-inch polar telescope may be seen as the tower jutting from the roof of the den, which was added to the house for that purpose

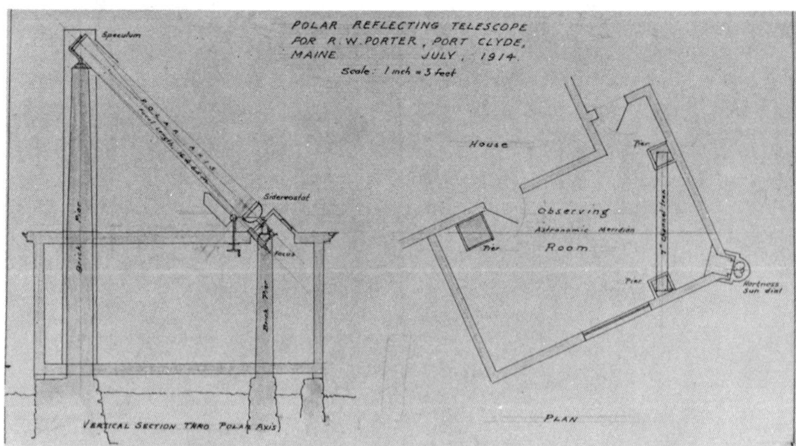

Original plan for Porter's 16-inch polar telescope

built one for this purpose. The mirrors were mounted on concrete pillars rising through the roof. The flat mirror caught the starlight and reflected it upward parallel to the axis of the earth to the paraboloid. At first the paraboloid was tilted in the Herschelian form of telescope so that the light might be focused past the flat, thus eliminating the need for a small diagonal mirror. But the scheme proved unsuccessful because of astigmatism in the image caused by the tilted mirror. A third mirror was then added, allowing the untilted mirror to focus excellent images. The eyepiece was mounted in the ceiling of the observing room, so Porter had to face south and look up at a forty-five degree angle, not the most comfortable position for long viewing. To keep a star in the eyepiece for long periods of time, the sixteen-inch flat mirror was driven by a mechanism powered with a falling weight. Setting circles were added as part of the essential design because no finder could easily be incorporated in this type of telescope. The entire telescope and den, as the new observing room was always called, cost about $150.00.

At last Porter had a powerful telescope and a heated room in which he could observe throughout the nights of the damp and cold Maine winters. The idea of a heated observing room was not new. Several professional telescopes had been built around this concept, and at least one amateur, that of James Hartness, who had recently completed a ten-inch refracting telescope (all optics by John Brashear) mounted on a turret of his own invention that turned to follow the stars while the observer remained comfortable inside.

Soon after completing his new den and becoming thoroughly convinced of its great potential to maintain amateur interest in serious observing, Porter decided to write an article describing his efforts and their rewards for *Popular Astronomy*, the same magazine in which he first read about making one's own telescope. Although the style of sketches on making telescopes that were to become familiar to thousands of amateurs years later did not appear here, Porter did have a drawing of his telescope and a photograph of his observatory. The manuscript, dated December 1915 from Land's End Observatory, Port Clyde, appeared in the May 1916 issue and became the forerunner of many articles on telescope making and astronomy. The same article was later reproduced in the *Scientific American Supplement* for August 4, 1917.

In this article he mentioned certain disadvantages to his type of telescope. For example, the region near the celestial pole could not be viewed since the concave mirror was in the way. An extra mirror was required above that used in the conventional "Newtonian" type of reflecting telescope, and condensation on the exposed mirrors was a problem. But Porter also wrote of the greater advantages. "Unquestionably the greatest gain lies in the amateur being able to work in a closed, comfortable room, which is as much a part of the home as his den or living room." The room could serve as a library. "With all his books and data at his elbow he will freely consult his references whenever a question arises at the eyepiece." The owner could proudly show his guests "the wonders of the universe with the same ease as showing him a beautiful painting in the library." One of the greatest conveniences was the fixed eyepiece. The observer could remain seated comfortably with all necessary controls within his reach. "The writer has found this stationary position of especial help in drawing the details of the Moon."[3]

Porter's telescope had one disadvantage that he never mentioned: the flat mirror was not large enough to allow the parabola to be completely filled with light except when observing near the celestial pole. For example, when observing an object on the celestial equator the flat mirror had to be tilted at a forty-five degree angle to the face of the parabola, causing the parabola to be partially bathed in an elliptical bundle of light. This was a small disadvantage, and one that could be easily overlooked.

No description of Porter's comfortable observatory is complete without mentioning the four-inch meridian telescope mounted in the window in the south corner of his den. He designed it to scan the meridian since he wanted to use it only to obtain local time. This was an outgrowth of his responsibilities of celestial time keeping in the Arctic. In later years this fascination nurtured his love of sundials and sunclocks.

The moon has always been the subject of intensive investigation by amateur astronomers. It is one of the few celestial bodies close enough to the earth to allow much detail to be seen. One can explore for hours and continually find mountain peaks, craterlets, and other markings unseen before. To Porter there was an added attraction that helped bond his past to the present. In his second article in *Popular Astronomy*, for October 1916, he showed a series of moonscape sketches and described how in his closed observing room he frequently "caught himself transported to that body, and, in imagination, viewing her scenery from some crater lip or the vast expanse of one of her sea floors." He could not help but be "struck by a strange likeness of the moon's general aspect to our own polar regions." He went on to describe the likenesses. "The long reaches of the frozen polar ocean, traversed by immense pressure ridges and tidal cracks, the dazzling whiteness and clear cut shadows, the desolation and loneliness"[4] all had their counterparts in the lunar appearance.

A second lunar project combined his artistic ability and astronomical curiosity. He labored many hours over a drawing of the full face of the moon, but not as it appears at full. Instead he chose to draw the moon as though each part were experiencing sunrise. This way long shadows would show the details of the mountains and craters. The single drawing based on Harvard photographs taken in Jamaica in 1900 was intended strictly for the amateur as an aid in identifying lunar details.

When it was completed he sent it to Hartness for comments. Hartness was overcome by the beauty and exactness of the drawing and by the fact that Porter had sent it to him. Recognizing it as a great aid to students and amateurs, he urged Porter to have it copyrighted. "I know your general make up about running in the opposite direction when you see a dollar," Hartness wrote on December 20, 1915, "but it strikes me that this is something that might bring a dollars and cents return."[5] There is no evidence that Porter ever pursued this, but it was copyrighted in the November 1916 issue of *Popular Astronomy*.

Porter was so happy with the performance of his heated observatory that he continued to spread the news. In the May 1917 issue of *Popular Astronomy* he published an article describing the various possible arrangements for making closed observatories with reflecting and refracting telescopes. But observing had not taken such hold of him that he had lost interest in the art of making the optical surfaces that allowed him to do the observing in the first place.

He set out to experiment with photographing the shadows that appeared on the mirror surfaces. Photographic records of the optical surfaces on telescope lenses, including the forty-inch of Yerkes Observatory, had been obtained earlier by J. Hartmann. Now Porter was the first person to try to do this with mirrors. His light source was an acetylene flame inside a brass tube. He placed the film plate behind the knife edge without any intervening lenses. One of the mirrors he used to make the photographs was one of the first he ever made, a ten-inch disc, which turned out to have a turned-down edge. He discovered that the photographs were able to show fine detail invisible to the eye on the mirror surface. In June 1918 he had an article on his work in the *Astrophysical Journal*. In appreciation of this effort, the French Astronomical Society offered him membership in their organization. But, in his modest way, it was an honor he decided to decline.

Porter was also fascinated with the mechanical design of telescopes. From his architectural studies at MIT he had acquired a sound knowledge of mechanical engineering, and in telescopes he discovered an excellent opportunity to apply it to a practical end. Telescope mountings provided a particular challenge. Not only do they have to support the optical elements free of vibration from the ground and wind, but they must also transmit a smooth and even motion to follow the stars. An early product of his design work was completed in 1917 and appeared the following year in *Popular Astronomy* (March 1918) and *Scientific American Supplement* (August 3, 1918). It was a proposal for a new form of mounting for large reflecting telescopes. The idea incorporated a yoke to hold the tube, thus allowing the center of mass of the entire telescope to fall within its three supports. Although this must be the first consideration for any flexure-free and vibrationless structure, it was not always so in earlier designs. Thus he demonstrated at an early stage that he understood well the principles involved in successful mechanical design. This simple but ideal suggestion had great significance, for years later it was incorporated in the design of the 200-inch telescope and had some influence in bringing Porter into the project, an amateur among professionals.

It has often been said between amateurs that Porter was the founder and patron saint of the amateur telescope making movement in the United States. While it is true that he did more than anyone else to inspire, guide, and instruct the amateurs that swelled the ranks after he moved to Springfield, Vermont, he was certainly not the founder of the movement and never claimed to be. In fact, in his first article in *Popular Astronomy* in 1916, he referred his readers to Professor M. T. Fullan of the Alabama Polytechnic Institute, Auburn, Alabama, who had recently organized an association of amateur telescope makers for the purpose of spreading the fever as well as instructions.

By the fall of 1915 the Porters had moved to 829 Beacon Street in Boston so Russell could teach architecture at MIT. Later they moved out of Boston to Newton. But they always returned to Port Clyde for school vacations and holidays. For three years Porter taught architecture as an Instructor in Design.

When the World War caught up with the United States, Porter received a request from James Hartness to go to Washington, D.C., to work at the National Bureau of Standards. The country was having trouble filling its needs for optical instruments and Hartness, influential in the machine tool industry, felt he knew just the man who could help. So for part of the last year of the war, the Porters lived in Washington while Russell contributed to the production of prisms and flat optical elements and experimented with silvering of mirrors. Alva H. Bennett, who was then a laboratory assistant at the Bureau, recalled that Porter "had a little room at the end of the hall about 8′ x 8′ in size and he kept the door closed most of the time. We did know that he was working on optical polishing and did silvering." The work was not without its hazards. "It seems I recall that the silvering solution exploded once and that the walls of the little room bore evidence (Brashear's Solution stains) for some years later." [6]

By early 1919 when Porter was no longer needed at the Bureau and was thinking of returning to Port Clyde, James Hartness again entered his life. He wanted Porter to come to Springfield to work for him on an optical invention he was trying to develop. His idea would give manufacturers a means of inspecting screw threads and thus improving quality control, which the war had shown to be decidedly lacking. Here was a chance to continue his optical hobby as a vocation—and with Hartness, who had helped him so generously already, as employer. Porter accepted the offer, and the family moved to the town of his birth.

Porter drawing of the home he rented at Hill Place, Springfield, Vermont

Russell, Alice, and Caroline Porter in Springfield, Vermont
  Date unknown

IX. RETURN TO SPRINGFIELD

His influence in the machine tool industry had put James Hartness on various national committees during World War I, one of which had been set up to study the problems of standardization and quality control of screw threads. Here he had learned that there was no method that would allow rapid and qualitative inspection of the threaded parts as they were manufactured. He had an idea that the required means of inspection might be made available by optically projecting the screw, greatly enlarged, onto a screen and comparing the image with a magnified profile of the desired piece. Any departures from the desired shape could then be seen, and the inspector could tell exactly what was wrong with the screw.

This idea of a projection lantern had been used for many years in the inspection of thread gauges and precision screws although it had been denied use for manufactured parts because of the time-consuming adjustments involved. Hartness's idea involved putting the threaded parts in a fixture that had already been adjusted for all the parts being made. It would not be necessary to adjust each one. In this way the reliability of the machines that depended upon these threads could be improved.

It was this challenge that brought Russell Porter to Springfield in the first part of 1919. Aged forty-seven and balding, he was embarking on a new career, one that would grant him the privilege of earning a living from his hobby, mechanical and optical tinkering. Hartness had known Porter for a long time, had seen him pursue his desire to be an arctic explorer, had helped and inspired him when he turned his drive to designing and building telescopes, and had decided that here was the man he needed to turn his idea into a working and useful contribution to the industrial community.

From the Jones and Lamson Machine Company Porter rented a house overlooking the town square at 2 Hill Place and settled his family. Then he set about his work for Hartness. The first problem was to transform the latter's idea for a projector, or *comparator* as it came to be named, into a working model to prove its feasibility. This required mechanical and optical design and tinkering. Porter was well prepared to tackle the mechanical work. He was also experienced in fabricating and testing mirrors. There was one thing, however, in which he had no experience: the designing of the projection lenses that were the very heart of the machine. To create a lens system capable of projecting a sharp and precise image of the screws required a thorough understanding of the optical principles involved and the application of these principles through many long hours of tedious calculations. Few people are inclined to pursue this half-art half-science, and Porter was not among them. To fill the gap Professor Charles S. Hastings of Yale was enlisted, and later other men helped on the work. Thus the project got rolling.

By the spring of 1920 Porter and the rest had succeeded in assembling a working model of the comparator. It was a rather large affair and needed refinement before it could be sold as a commercial product. But it proved out the idea. The life of every person in the town was influenced by the success or failure of the town's machine tool businesses, and everyone was eager to learn of any new product that might bring good fortune his way.

Porter wrote a news story describing the emerging product for the townsfolk. "The projection lantern has for years been used to cast profile shadows of opaque objects on a screen for the purpose of studying their outlines." He went on to explain how this principle had been modified so that the screw could be held and manipulated in the machine so as to reveal on the projection screen its various elements. Lead errors, diameter errors, errors due to roughness, out of roundness, drunken threads, bent screws, all were open for inspection. He added, "The chief merit of the Hartness Comparator lies in its ability to show on a chart one of the threads of a screw with all these elements in plain view, greatly enlarged."[1]

Because this was Hartness's first venture with a commercial optical product, the company had no facilities or experience in working the optical parts. Porter was put in charge of setting up an optical shop and training the workers in the skill of grinding and polishing the lenses. The success of this early endeavor would grow through the years until it became one of the Jones and Lamson Machine Company's major products in its line of machine tools.

While working for Hartness, Porter did not lose interest in telescope making and astronomy. He inspired a group of townsfolk, mostly connected with the shops, to make their own telescopes, and he spent much of his free time designing new mountings for telescopes. Meanwhile, his daily life continued to reflect its healthy balance between outdoor adventure, scientific and engineering studies, and the arts. Now nearly fifty, he still sought the out-of-doors, whether it was to paint a Vermont scene or chase an escapade.

In the first year of his return to Springfield, 1919, a large Fourth of July celebration was planned. As part of the festivities an aviator was to come up from Swampscott, Massachusetts, to give a demonstration of aeroacrobatics, partly to spread interest in aviation to Vermont. If the pilot and his mechanic could arrive in time, the public could also take rides of a more sedate sort. After a great deal of trouble with visibility, losing the way, finding places to land for gasoline, and avoiding the hills of southern New Hampshire, they finally brought the plane down in time to give the patient crowd an eyefilling display of daring stunts—loop-the-loops, barrel rolls, and nose and tail spins. The following morning the promised rides were available, and Porter was the first person to step forth. After returning safely to the ground, he told the local newspaper reporter that it was his impression that the air had no greater terrors to offer than the Arctic. Very soon afterward, the plane was caught in a downdraft and crashed on the field. Miraculously, neither passenger nor pilot was seriously injured.

Often Porter would load Alice, Caroline, a picnic, and his art supplies into the Ford and drive out to spend the day in some interesting natural setting. A favorite outing was up Breezy Hill to the old family cabin, "The Shack." With his sketch pad or pastels Porter recorded the uniqueness of the world around him. If he was a prolific artist all his life, he also was a modest man who rarely exhibited his paintings and sketches. In the spring of 1920 he did show six water colors of Eskimos and arctic scenes at an exhibit of Vermont painters.

Porter also took his family on camping trips. He got a great deal of pleasure out of teaching Caroline woodlore and how to get along in the out-of-doors. They visited old mines and learned about rocks. Alice was a good sport and went along on these trips, but she was not as enthusiastic as her husband or daughter.

During the summers the family made occasional trips to Port Clyde, where Porter still owned his home at Land's End. One year at least Hartness gave Porter four weeks off so he could spend a month in Maine.

In October of 1921, Alice and Russell drove to Mount Mansfield

James Hartness, left, and Russell Porter in Springfield, Vermont, in September 1920

in the northwest part of Vermont with Governor Hartness (as their old friend had become in the election of 1920) and a few other guests for a camping trip. Some of the party drove up the highest mountain in the state and camped overnight in tents. This was one of the occasions when Porter was called upon to entertain the group with stories from his arctic memories. Although he was a reserved man who never offered to talk about himself, he enjoyed telling about his arctic adventures when coaxed into it.

One of the neighborhood lads and a friend of Caroline, John Pierce, grew up with the lasting impression that while at home Porter was almost always working on something. What little recreation he took he took seriously. John recalled how one winter day Porter decided to build an igloo on the front lawn for Caroline and her school friends. The snow was just hard enough for cutting blocks, having been piled up from the walks. He used an ordinary carpenter's hand saw instead of a snow knife. Explaining to the wide-eyed and excited children exactly what he was doing, he placed the blocks in a spiraling circle, each holding the other until the igloo finally domed over. The children probably helped pass blocks to Porter as he worked from the inside, and finally one of them dropped the keystone block in place. Suddenly all activity ceased. But soon the tip of the saw came out the side, and Caroline's father crawled out the freshly cut hole. Then he made the tunnel and left the children to play with their new ice palace.

Porter always enjoyed playing with celestial navigation. In 1920 he wrote an article, "Latitude without Instruments," which appeared in *Popular Astronomy*. In it he demonstrated how it was possible to measure one's latitude with the simplest of implements, things one might find around the house or even on a camping trip. With a plumb bob nailed up in a tree, a steel tape measure, and sighting aperture, he measured the angle between the zenith and the sun as it crossed the meridian at midday. This would be equal to the latitude were the sun at the equinox. By referring to solar tables in *The Nautical Almanac* and adjusting for the sun's declination, he was able to determine his latitude to within a minute of arc as checked by a theodolite.

A man who shared Porter's enthusiasm for the out-of-doors was Ralph Flanders, manager of the Jones and Lamson Machine Company since 1914 and later the senator from Vermont. Each year while

Porter water color of Mount Katahdin, Maine, from Chimney Pond

the Porters lived in Springfield, Ralph and Russell took a camping trip somewhere into Quebec north of the St. Lawrence River. Years later Senator Flanders told the author that he could not remember how many trips he and Porter had taken during the 1920s, but he did recall one that made a lasting impression. It was before the days of widespread use of oil-company road maps, and a traveler had to rely on intuition and what information he could pick up along the way. One night they were camped on the land of a French Canadian farmer—somewhere in Quebec. Flanders asked Porter, "Where are we?" That was enough to make Porter reply, "I'll tell you."[2] First he brought out a saucer, set it on the fender of the car, and poured into it some maple syrup they had along for the pancake mix. Then he brought out his sextant and took sight on the north star, bringing together the direct image of the star and the reflected image in the maple syrup. The syrup acted as an artificial horizon and made a fine substitute for the real horizon at sea. The altitude of the north star was, of course, half the angle read from the sextant. By making the proper correction for the star's not being exactly on the north celestial pole, the altitude was converted to their latitude.

Longitude was not so easy. This required the accurate knowledge of Greenwich time. But trusting the accuracy of his watch and knowing the longitude at which he last set it for local time, Porter was able to overcome the lack of Greenwich time. He pulled from under the seat of the car *The Nautical Almanac*. Selecting a star in the eastern (or it might have been western) side of the heavens and using the artificial horizon of the maple syrup again, he measured its altitude. Finally, using this, the almanac, other tables, and his watch, Porter determined their longitude. He scribbled the longitude and latitude on a piece of paper, handed it to Flanders, and said, "That's where we are."[3]

These two outdoorsmen made one of their trips together in 1924, this time to climb Katahdin, at 5,267 feet the highest mountain in Maine. Flanders recalled they drove up to Stacyville to the east of the mountain. In 1924 the mountain was part of a state game preserve, there were few direct routes to its base, and Mr. Baxter had not thought of a state park. The usual way was to approach from the south to take advantage of a privately owned logging road to Kidney Pond and then climb the Hunt Trail. Flanders said, "We took the hard way, I don't know just why, but it seemed to be the natural way with

Russell." They made inquiries for a guide, and a man working on a road construction job was recommended. "We put the proposition to him and in Russell's language he left his pick axe hanging in the air and came with us." [4]

That evening, on the East Branch of the Penobscot River, they learned about hunting-lodge etiquette. Having finished the first course of supper, they patiently waited for the waitress to clear away the dishes. Instead she asked if they were through. When they replied yes, she asked why they had not stacked their plates. And so they stacked.

The next morning burdened with heavy packs they started for the foot of the mountain. Before reaching the boundary of the game preserve the guide shot spruce grouse, which were good eating. Then he hid his rifle, and they went in to meet Warden LeRoy Dudley, who gave them the key to the warden's camp at Chimney Pond.

Once at Chimney Pond the weary climbers were able to rest, surrounded by the most spectacular scenery in all of Maine. The camp was set neatly on the bank of the pond whose existence depended on a moraine. Just beyond the pond, the mountain begins to rise, slowly at first and then more rapidly. The eye struggles to grasp its entirety as it continues to soar over 2,000 feet from pond to summit. On its way to the sky the mountain mass captures over half of the horizon, and one must lift his eyes and turn his head from side to side to follow the complete outline, for he is standing in a great basin carved by the weight of ice from eons past. Towering buttresses put their shoulders to the mountain. High up on the horizon a jagged knife-edge stands ready to slice at clouds as they roll in from the west and struggle with the air currents rising from the mammoth amphitheater.

In such surroundings the men might have been tempted to forget climbing and be content with looking. But climb the mountain they did. It was fall, and after returning to Chimney Pond they were snowed in for two or three days before being able to hike back to civilization.

These camping trips offered something to Porter that Springfield failed to give. In them he found a chance to shake off the confines of a neighborhood and keep alive the memory of more youthful adventures. In some respects the town held little interest for the Porters. Their friends seemed to be more widely scattered. One of the neighbors, John M. Pierce, the father of Caroline's friend John and teacher of manual arts in the high school, did become a close friend of Russell, however. Pierce had caught the fever of making telescopes from Porter, and the two men spent many hours huddled together over the problems of shaping glass and metal into a new telescope.

Porter was not, and was not thought to be, a snob. He was interested in other people, always ready to listen to what they had to say, and ready to carry on an interesting conversation. But he was not a joiner, not socially or civically minded. The few town affairs that did interest him had a direct connection with his own skills. For example, he designed the town seal, and soon after returning to Springfield in 1919 he did a series of eight sketches for a book about town folklore.

At home Porter had a room upstairs where he kept a drawing board and other drawing materials. Here he could work on designs for the particular instrument he happened to be thinking about at the time. His daughter recalled that he kept drawing pads all over the house, always ready for sketching a new idea for a telescope or drawing a scene that caught his eye. There were never any idle moments.

Porter was very quiet about the house. He seldom raised his voice and was slow to anger. But everyone else had to shout to overcome his deafness, his scar from the Arctic. His close friend Oscar S. Marshall, a journeyman machinist and foreman of the Special Tool Department at the Jones and Lamson Machine Company, once wrote that he never heard Porter complain about his deafness. Porter could not stand complaining by anyone. Strangely, while driving a car, he could hear the passenger beside him talking in a normal tone of voice. Another local man remembered Porter as a poor driver because he was hard of hearing and his mind wandered, but Porter drove across the country many times without recording a single accident. His favorite automobile was the Model T Ford; he never wanted anything else.

During his arctic years Porter had smoked a pipe. Then for some unknown reason he switched to cigars. Although he was a captivating storyteller, his language always simple, his listeners might have to endure the fumes of a long black Pittsburgh stogie. He loved his cigars. He would inhale the smoke, then exhale it and then inhale the same puff again, this time through his nostrils, and finally exhale the whole thing. These instruments of torture were a never-ending source

Drawing by Porter of Vermont maple trees, 1920

of amazement to his friends. In his later years, he switched back to a pipe.

Porter was not practical enough to be concerned about his personal appearance. Usually Alice had to remind him to change his clothes when they were going somewhere. He was more interested in being comfortable in his knickers.

It is not possible to assign one great passion to Porter, he was interested in too many pursuits. However, near the top of the list must be placed working glass, that is, grinding and polishing it to the fine degree of perfection required of optical instruments, especially parabolic mirrors. To the very end of his life he had a strong urge to do things with his hands. Whenever he found himself working at an excess on other projects, he longed to get back to a piece of glass, to feel it take shape under his guidance, to push it back and forth over the rouge slurry and pitch lap with the strokes that had become second nature after countless hours of patient practice. Then there was the joy born anew every time the testing showed the mirror to be performing as desired.

Those who considered Porter skilled in making mirror surfaces for telescopes surely must have considered him skilled at many other talents as well. Hartness called him his Leonardo DaVinci. An exaggeration, but when the need arose, Porter could be a draftsman; his thousands of sketches, water colors and pastels may be the greatest product of his dexterous skills; he had been an architect and navigator and was still a practicing amateur astronomer; he even had an interest in music and had experimented at composing quartets, some based on primitive music he had heard the Eskimos dancing to in Greenland; and he was a teacher of great patience and inspiration.

Always Porter was the type of man whose mind was constantly working up new ideas, no matter what the time or circumstances. Once, for example, he and Oscar Marshall climbed up Mount Ascutney, a few miles north of Springfield, and near the top bivouacked for the night in a dry brook bed. But Marshall having caught the germ of telescope making from Porter, the two men did not put the night to bed. They sat by the fire and talked about a design for a big astronomical telescope. Neither man had thought to bring along pencil or paper, so Porter improvised—with birch bark and charred sticks. The scene was lighted by a good pine knot with a stem.

Porter's imagination took hold of the charred sticks, and before the
night receded it had created the idea for a large reflecting telescope.
Marshall later wrote it resembled the turret telescope built several
years later on a hilltop near town. But telescopes are the story of the
next chapter.

Vermont covered bridge, Porter water color
  Date unknown

X. STELLAFANE IS BORN

The following persons have signified their intentions of making a reflecting telescope for their own use, under the direction of Russell W. Porter, and they are hereby requested to meet in the Foremen's Room of the Jones and Lamson Mch. Co., Tuesday, Aug. 17, 1920 at 11:30 A.M. standard time, for preliminary arrangements preparatory to beginning work. Mr. Porter will be present to preside.

Names:	Glass size
Russell W. Porter, founder	16"
Prof. John M. Pierce of the Springfield Co-Op Course	10"
Oscar S. Marshall	4"
Oscar P. Fullam	8"
Everett H. Redfield	6"
Frank H. Whitney	7"
Ernest N. Brookings	7"
Raymond P. W. Fairbanks	4"
Roy J. Lyon	8"
Albert A. Herrick	8"
Gladys A. Piper	5"
Ralph A. Baker	9"
Guy E. Baker	9"
F. Eugene Lockwood	6"
Clyde P. Baldwin	6"
Charles A. Longe	4"
Carlton B. Damon	9"

Porter polishing a telescope mirror, circa 1920

Porter's first class of telescope makers, October 1920
Left to right: Oscar S. Marshall, Clyde P. Baldwin, F. Eugene Lockwood, Russell W. Porter, Raymond P. W. Fairbanks, Gladys A. Piper, Oscar P. Fullam, Ralph A. Baker, Frank H. Whitney, Roy J. Lyon, Everett H. Redfield, Carlton B. Damon, Ernest N. Brookings, Guy E. Baker. Missing: Albert A. Herrick, Charles A. Longe, John M. Pierce

The above persons will be considered charter members of this association, which may develop into a permanent enterprise for the making of optical devices of various forms. Mr. Porter has placed an order for the glass discs covering the required sizes as per the above list, also the necessary pitch and grinding mediums. Working rooms are to be provided for members in the southwest corner, lower floor, of the brick shop occupied by the Automatic Die Division.

No obligations are incurred by any member thus far beyond the cost of their discs and the grinding materials purchased for their use, and lighting of room and any other things which by mutual arrangement are agreed upon.

Please be present at the meeting.

Signed:

Oscar S. Marshall[1]

Thus Russell Porter brought together the first group of men and one woman in Springfield to put their minds and hands to telescope making. Wanting to share with others the satisfaction of making and using one's own scientific instrument, he set about teaching these people. Arrangements had been made with Hartness, always willing to help the enthusiastic amateur, to use an empty room at the Jones and Lamson Machine Company and, when the time came, to use some of the machining facilities. Perhaps the facilities were available because of a recent downturn in the economy that had severely hit the machine tool industry. Most of the men were employees of Jones and Lamson, on a shortened work week, and in need of a time-consuming hobby.

First Porter instructed his class in the technique of mirror making. When told they would have to make the mirror surfaces a thousand times more precise than the parts they were used to in their daily work, they readily accepted the challenge. They took the glass mirror blanks and carefully worked them back and forth over other glass discs until the grinding abrasives and water in between shaped the discs to matching spheres. Porter kept a watchful eye on their operations to see that each mirror was made concave to just the right depth. Then they spent many hours grinding with finer and finer abrasives until their mirrors were almost smooth enough to reflect an image. The next step was polishing, which they did by strok-

Oscar Fullam and his wooden telescope, circa 1921

Oscar Marshall and one of his telescopes, circa 1921

ing the mirror over a tool of pitch covered with a slurry of very fine powder called rouge. The polishing process took many hours, but at last the mirror surfaces were smooth and nearly spherical.

Porter demonstrated to his pupils that in order to have their mirrors work properly in their telescopes, they must refigure them from spheres to parabolas of revolution, or paraboloids. He taught them the knife-edge test, invented by Leon Foucault during the nineteenth century to examine the shape of mirror surfaces. This remarkably simple test, which required only a pinhole of light (to be reflected from the mirror surface) and a knife-edge (to intercept the rays of light as they returned to the eye in a sharp focus), created quite an impression on the reserved Vermont craftsmen. Never had they seen such a basic apparatus that could detect surface irregularities as small as a fraction of a wavelength of light. If the test apparatus was simple in principle, a certain amount of skill was required to understand the meaning of the shadows that appeared on the mirror surface. It was this skill that Porter brought to the group, helping them interpret the shadows and showing them the proper strokes to be used to bring their mirrors to the desired shape.

Into the early months of that winter, work on the mirrors progressed two or three nights a week. All but one man finished. Then Porter showed the group what was needed in the rest of the telescope: the equatorial mounting, tripod, eyepiece holder, and all the other parts. Each was encouraged to design his own telescope, using his own best skills. The pattern maker made his of wood. The machinists crafted metal parts. No two products were alike, and yet they were the equal of many professional telescopes both for precision and workmanship.

The finished telescopes ranged from four to ten inches in aperture and were capable of yielding magnifications up to at least 200 power. Other than the mirror blanks and grinding abrasives, only the prisms and eyepieces were purchased, a satisfying accomplishment in a time of thin budgets.

While Porter was instructing his class of mirror makers, he was also designing a new type of telescope mount. This project grew out of his continuing desire to move observational astronomy from the realm of casual and passing interest, usually caused by physical discomfort, to the sphere of serious study made possible by ideal conditions. Earlier, at Port Clyde, he had experimented successfully with

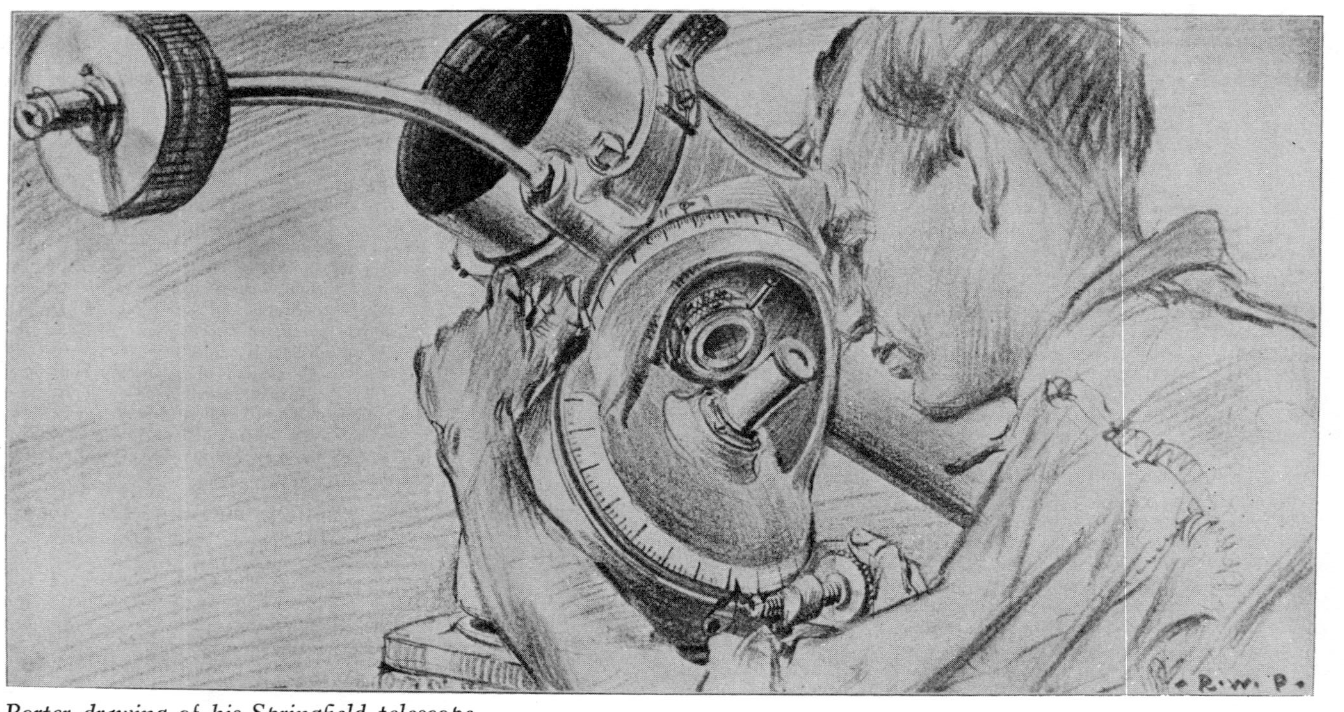

Porter drawing of his Springfield telescope

a closed observatory, but this technique was beyond the means of the average amateur. The next best thing would be a fixed eyepiece. Then, even if the observer is exposed to the cold and wind, he can sit or stand in the same comfortable position instead of twisting and turning in all kinds of awkward positions to follow the eyepiece as he points the telescope about the sky.

In 1920 Oscar Marshall made a prototype of Porter's invention, a fixed-eyepiece telescope. By starting with the normal Newtonian telescope and adding a third reflection, the light was made to travel parallel to the declination (or north-south pointing) axis and then up alongside the polar axis to the eyepiece. The observer could stand in a comfortable position and observe all parts of the sky while the eyepiece moved only a small distance. The principle is similar to that of the coudé focus used on large telescopes to direct the starlight down the polar axis to a fixed spectrograph that is too large to be mounted directly on the telescope.

In honor of his home town, Porter decided to name his new mounting the Springfield Mount. In designing it, he had applied the soundest of engineering principles to eliminate the flexure and vibration that had been prevalent in earlier telescopes. The bearing surfaces were given large diameters and were held together with central studs, preventing vibration and shake in the bearings. The parts were massive, and the center of mass of the entire mount and telescope was brought to the intersection point of the two axes to remove flexure. Simple tangent arms were used to allow the amateur to construct inexpensive fine-adjustment and drive screws on the axes.

In the final design the starlight was made to travel directly along the axes so the eyepiece remained perfectly fixed. The telescope was manifested in aluminum castings. Even though the end product was a sturdy mounting that gave vibration-free star images, the required castings and machining discouraged many amateurs from incorporating it in their telescopes. Those who did, however, were well rewarded for their efforts; they had telescopes that were the equal of any for sturdiness and allowed many hours of continuous, comfortable observing. Some amateurs would later demonstrate their ingenuity by making the Springfield Mount from easily available pipe parts, although the resulting instrument was not as sturdy as the original.

Oscar Marshall and his prototype of Porter's Springfield telescope, circa 1921

Porter's telescope makers after an all-night observing session on Mt. Ephraim, Springfield, Vermont, circa 1921

Porter's telescope makers placing a bronze survey tablet on Hawks Mountain in 1922. Porter is holding the tablet.

The telescope makers at Hawks Mountain, Springfield, Vermont, examining a prototype of Porter's Garden Telescope, circa 1921

Eager to share the excitement he had experienced as an explorer, Porter took his group out for observing sessions. Sometimes they met in backyards, and sometimes they took weekend camping trips to find dark sky and open horizons. Imagine their enthusiasm, setting out with instruments powerful enough to explore the universe, made with their own hands, to share the wonder and awe of the early astronomers. They explored the wastelands of the universe as their instructor had explored the wastelands of the Arctic. Past the sun, past the comets and the asteroids, past the planets to the stars and beyond, these ordinary mortals wandered with their self-made instruments. Together they had a miraculous experience, yet it was not a shared experience. To each one, astronomy meant different things, and slowly they discovered that what they were searching for was not the stars, but something hidden deep within themselves. They were true explorers.

Sometimes Porter and the telescope makers drove to the open fields on the slopes of Hawks Mountain just north of town. Here were the open spaces needed for excellent observing and fine camping. Sometimes their families came along. One day in September 1921, they climbed to the top of Hawks Mountain and reoccupied an old surveying site that had been used in 1875 as a station in the primary triangulation of the geodetic survey of the United States. Because Porter felt the location of the village of Springfield was not accurately known, he took the longitude and latitude from this survey site and with the aid of a theodolite and his telescope friends extended the triangulation to the village. A year later they camped here again, erected a tower cut from small trees, and dedicated a bronze tablet explaining what they had done.

Another time they made an all-night camp on the summit of Mount Ephraim, a large, rocky, wind-blown knob south of the village. About fifteen men with two telescopes sat up from dusk till dawn exploring the heavens. There was no time for sleep; one might miss a celestial object new to him or some argument about planetary motions or stellar evolution.

These were men of simple backgrounds, some with schooling limited to rural high schools. Yet from an inspiring teacher they were learning much about astronomy. Greenwich time, celestial equator, ecliptic, right ascension, all became familiar, and they soon understood the motions of the celestial bodies. They continued to learn

Some of the instruments made by Porter's telescope makers as exhibited at a county fair, circa 1922

about optical instruments, how to make spherometers, and how to measure precisely the curvature of telescope mirrors. Porter explained to them how the interference of light was used to calibrate the gauge blocks they used in their trade and how it could be used to measure the diameters of stars.

The amateurs had been meeting on an informal schedule but now it was agreed that a club should be formed. On the evening of December 7, 1923, the first official meeting of the Springfield Telescope Makers was held at the home of Oscar Marshall, who was elected secretary-treasurer. Porter was elected president, and John Pierce, vice-president. An article was written into the constitution stating that to become a member a person must first make a speculum (mirror). Although Gladys Piper had successfully finished her mirror, she had moved away in 1921, thus for many years, the club was exclusively masculine.

The most important discussion during that first meeting concerned the construction of the bungalow that had been started that fall. Backyard observing had not been conducive to the sharing of ideas and experiences, and they had grown tired of carting their telescopes, camping equipment, and families to observing sites on weekends.

Porter had inherited the land on top of Breezy Hill west of town where he had taken Anthony Fiala to The Shack, back in 1906. Now he volunteered to lease it to the club. (It was finally sold to the Springfield Telescope Makers in 1928 for one dollar.) His experience in building cottages on the Maine coast came in handy both in design and in construction. To get things started he contributed $314.35 for building materials. The men trucked the lumber and other essentials up the hill, and everyone pitched in and did a share of the work in order to have a place to come in and get warm during long nights of observing.

Before the building was completed the amateur astronomers discovered that more money was needed. The extra funds could not be raised among them, so Porter headed up a small delegation and paid a visit to a certain friend in Springfield. The friend gave Porter a check, which without a glance he put into his pocket—with his typical lack of concern for financial matters. The check was more than enough to pay for the entire lumber bill.

It has not been recorded who this friend was, but almost cer-

A Porter sketch of the clubhouse built on Breezy Hill, soon to be called Stellafane

tainly it was James Hartness. After all, who had built his own amateur observatory? Who had helped Porter financially with his telescope building efforts in Port Clyde? Who had made available the facilities in the Jones and Lamson buildings? Also, the records[2] show that Hartness was made an honorary member of the club in December 1923 by a generous gift of $100.00 and later when the treasury was all but depleted, he matched this gift. On yet another occasion, when plans were being made to build a large telescope in front of the clubhouse, Hartness donated another $100.00

The interior of the clubhouse was finished in wood stained a light gray and resembled the Port Clyde cottages. The main room had a large brick fireplace and a low beamed ceiling. Some say the beams were curved downward at the ends to make the room look like a ship's cabin, a product of Porter's vivid memory of many months spent in the Arctic. Shelves were built into the walls of the main room and a library of astronomical and related books was gradually accumulated. Many of the books later were contributed by the *Scientific American* staff and the Harvard Observatory. Photographs of celestial objects, telescopes, and famous astronomers were put around the walls to add to the atmosphere.

By January 1924, the main part of the house was completed. At the January 30 meeting Porter suggested it be named "Stellar Fane," meaning "shrine to the stars." It was unanimously adopted. Before very long, however, the name was contracted to "Stellafane."

The little clubhouse on the top of Breezy Hill began its silent watch over the many years of astronomical activity. It faced north and looked out over the fields, far down into the valley where the wheels of precision machine tools were turning, and up again to the north to the rising slopes of Mount Ascutney on the horizon. On the north gable, a member carved the words "The Heavens Declare the Glory of God."

Porter was not an active church attendant. His deafness made it impossible to hear sermons. One Sunday morning when he and Oscar Marshall were heading for the clubhouse, they were approached by a deacon and asked whether they were going to church. Porter replied they were going to Breezy Hill and would not make any noise that would disturb him. He also reminded the deacon of the inscription on the gable of their temple to the stars.

By the summer of 1924 a small kitchen on the south side of

Porter, left, sketching at Stellafane, circa 1923

Porter's dream of a reflecting turret telescope, 1920, water color. (What actually was built appeared at Stellafane in 1930.)

the building was completed. Food could be prepared over a wood stove, suppers before meetings or snacks anytime during the long nights of observing. Upstairs two bunkrooms were tucked away under the steeply gabled roof, reached by folding stairs from the main room.

Even today the story is told of how Stellafane came to be painted a light pink. Working with a tight budget and eager to uphold the image of Yankee frugality, the members agreed to accept a gift of paint offered by a local merchant provided they did not specify the color. It turned out to be pink, and Stellafane remains pink to this day.

During the summer of 1924 Porter's telescope makers attracted the interest of Webb Waldron, a writer for *Collier's Weekly* and the *Century Magazine*. Mr. and Mrs. Waldron spent a July weekend at Stellafane as guests of the amateur telescope makers and astronomers. After a sleepless night of observing both the stars and the men's enthusiasm for their hobby, Waldron had enough information for two articles, which appeared in these magazines. In reading them one can see how impressed he was with this group of happy men.

Most were simple men; some had been members of the farm-to-factory migration. They were part of the industrial revolution, caught up in a way of life that suppressed individualism and creativeness. Their Yankee ingenuity, so often alluded to in discussions of Vermont virtues, had been stifled.

Russell Porter had opened up a whole new world to these men. Imagine a man working in a drab machine tool shop in the valley during the week and climbing the hill on the weekend to explore the universe among his friends. It was one of Porter's greatest satisfactions, knowing that he had inspired these men to a new plateau of living and understanding of life. This made him a "really happy man," as Webb Waldron called him, and his pleasure would grow when the hobby of telescope making spread far beyond Stellafane in later years.

As the word got around that some men were building their own telescopes, studying the stars, and actually having fun all the while, interest began to spread. New members joined the club, and their social status in the valley did not matter. At Stellafane all ranks disappeared in the common pursuit of scientific instrument making and using.

James Hartness, governor of Vermont in 1921 and 1922, generously supported the club but was not an active member. Ralph E. Flanders, manager of the Jones and Lamson Machine Company since 1914, and later its president, had been an active member from the formal beginning of the club. Ernest V. Flanders joined at the same time. He was a gifted and valuable engineer who worked closely with his older brother Ralph on new concepts in thread grinding that led to the Jones and Lamson automatic thread grinder in the mid 1920s. In later years Joseph B. Johnson, before serving as governor of the state, caught the telescope bug and has never lost interest in the club. Great men, simple men, leaders, followers, thinkers, tinkerers, they shared skills, exchanged knowledge, labored together, and took pleasure in one another's company.

Through the winter months the men often met at members' homes as well as at Stellafane. On these occasions the ladies were invited to provide refreshments, of course, but they were rarely invited to Stellafane. Sometimes, however, the restrictions were dropped, and Caroline Porter accompanied her father to the clubhouse supper and meeting.

During the first week in June 1925, R. M. Wilson of the U.S. Geological Survey and his assistant, R. W. Walton, as part of their summer work of primary triangulation to help complete the geological survey work in northern New England, climbed to the survey point on Hawks Mountain. Discovering the tower and bronze tablet put there by Porter and his friends, they drove to Springfield and inquired about the Springfield Telescope Makers, whereupon they were directed to Stellafane. The two men arrived late in the afternoon just as Porter, Fullam, Redfield, and Pierce were eating their evening meal. "Mr. Wilson introduced himself and it was not long before the sharp axes of the two veteran surveyors had limbed out all obstructions to an intimate acquaintance. Both Porter and Wilson have done work in Alaskan territory, and this alone was sufficient to establish mutual relationships."[3]

Wilson returned to Stellafane on the following Saturday, June 13, and was on hand for the meeting. During the day he made numerous observations, using Stellafane as the second of three triangulation points for his survey work. Much to his surprise during his first visit to Stellafane, he had discovered a transit telescope jutting from the west wall of the livingroom. Porter had erected it not too long before to time the passage of stars across the meridian, an idea that grew from his time-keeping measurements in the Arctic. Now Wilson used it to obtain data from the stars to establish the local longitude.

The following Friday, June 19, Porter and Marshall helped Wilson occupy the third triangulation point, Mount Ephraim, near Springfield. Porter was reliving his old surveying days as they helped Wilson record angles to various mountains in Vermont and New Hampshire. After the observations were taken, a bronze marker was set in the ledge.

Another visitor came to Stellafane the weekend of June 12–14. He was personally interested in making telescopes, and his visit was to have a profound influence on the coming growth of the telescope making hobby. His name was Albert G. Ingalls, and he was an editor from the *Scientific American*. Struggling with his own telescope mirror, he had become aware of the awakening interest in amateur telescope making. Thus he came on assignment to gather material for an article on how amateurs could build reflecting telescopes.

Friday evening Mr. Ingalls was a guest of the club at the home of Oscar Marshall. The sky was clear, and they enjoyed views of Jupiter, Saturn, star clusters, double stars, and nebulae through the host's eight-inch reflecting telescope.

The following afternoon and evening were spent at Stellafane, where several of the members had their telescopes set up. The sky cooperated, and the men spent the evening talking and observing. Perhaps the club members were just a little proud to be able to show off their finely crafted instruments and talk telescope making to an editor of a national scientific magazine.

By the time Ingalls left on Sunday morning he had learned much, how each man had studied and learned the skills necessary to fabricate a mirror, how each man had applied his own character to solve the many problems that came up. He observed how each of the machinists had applied the skills of his vocation and discovered an outlet for his ingenuity by creating mechanical things both functional and beautiful. The gleanings from that fruitful visit became the basis for an article in *Scientific American*. It was designed to discover how much interest in telescope making already existed and how much could be created among its readers.

Porter was always humble when he talked about himself. But such stimulus as helping plan this article could make him completely unselfconscious. Mrs. Ingalls enjoyed the memory of her husband and

Albert G. Ingalls during his first visit to Stellafane
Sketch by Porter

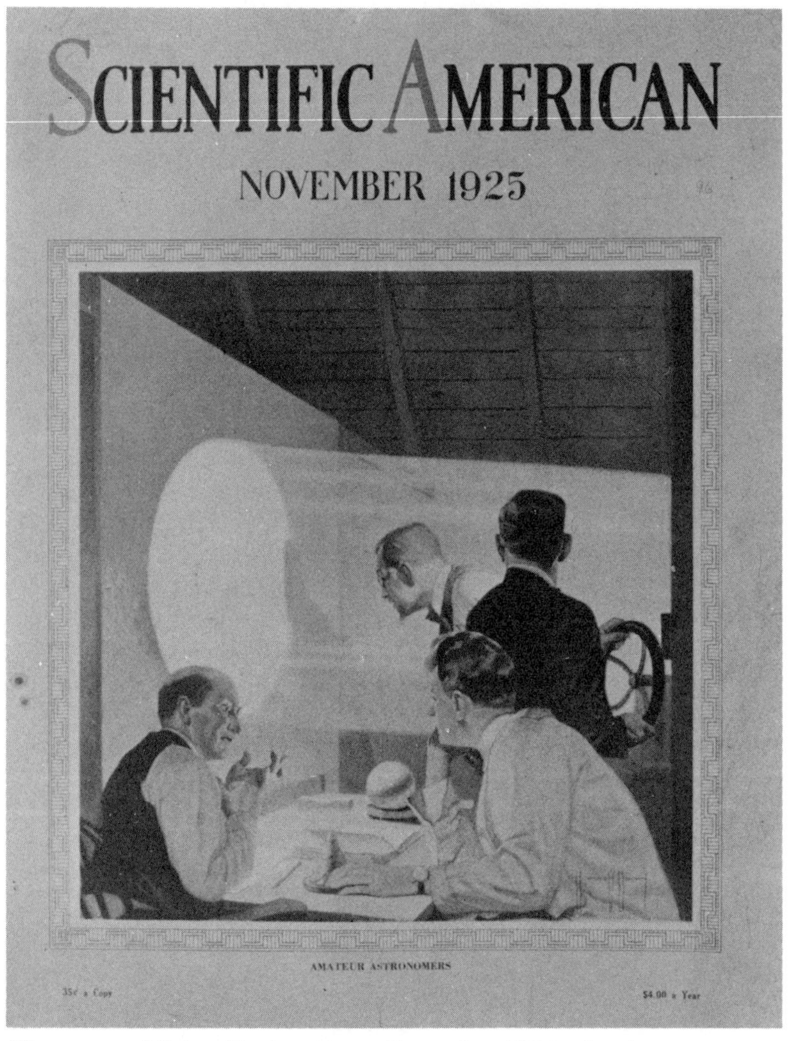

The cover of Scientific American, *November 1925, showing amateurs at Stellafane using their solar telescope*

SCIENTIFIC AMERICAN

THE MONTHLY JOURNAL OF PRACTICAL INFORMATION·

NEW YORK, NOVEMBER, 1925

"The Heavens Declare the Glory of God"

How a Group of Enthusiasts Learned to Make Telescopes and Became Amateur Astronomers

By Albert G. Ingalls

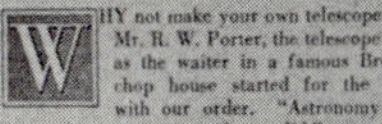

"WHY not make your own telescope?" said Mr. R. W. Porter, the telescope maker, as the waiter in a famous Broadway chop house started for the kitchen with our order. "Astronomy would mean a lot more to you if you did."

We had met to talk about Porter's hobby, astronomy. I had already heard quite a lot about this versatile man whose whole life had centered about the study of the stars. In his earlier years he had spent a dozen winters in the Arctic as astronomer, topographer and artist. Three years he had been with Peary, three more with Fiala in Franz Josef Land, and two years with Cook, who Porter says certainly did not climb Mt. McKinley. Other seasons he spent in northwestern Canada and in unknown Labrador. During all these years in the Far North where the Arctic stars fairly snap in the cold, clear air, he was studying astronomy.

Now he had settled down in the picturesque manufacturing village of Springfield, Vermont, tucked away in a deep valley in the foothills of the Green Mountains, where, as everyone in the mechanical industry knows, a famous type of flat turret lathe is made. Here he had fired a score of men with his own keen enthusiasm for the stars and had organized them into a group which is perhaps unique—machinists by day, amateur astronomers by night.

"You'd have no trouble in making a good telescope," he assured me.

"I could make the mounting all right," I replied, "but when it came to making the optical parts I'd be out of it. Only a handful of men in the world are skilled enough to do that fine work."

"You come up to Springfield, where I live," he laughed, "and I'll show you a good many home-made telescopes, made in spare time by men who knew nothing about it when they began. They'll tell you how any amateur—even an editor—can make his own telescope for less than fifty dollars, providing he's reasonably handy and will take pains. And it will be a real telescope, fit for serious work, not just a toy or a makeshift."

The "Poor Man's Telescope"

He went on to tell me how in the Vermont village a group of men, most of them mechanics in the local machine shops, had banded together to study the stars; how each one had made and mounted his concave mirror; how they had later pooled their efforts and built a sort of combined clubhouse, lodge and observatory on the top of a mountain near their homes. Here they gathered when the week's work was done, to study the stars. "The Telescope Makers of Springfield," they call their club, and none may join who has not made his own telescope.

When summer rolled around, I went to Springfield, as Porter had suggested, and there the amateur astronomers told me how they had learned their new avocation.

There are two common types of telescopes, the refractor and the reflector. The refractor is the ordinary type that everyone knows. It is like a big spyglass; you look *through* it, the light actually passing through its lenses. For serious amateur work such a telescope, having an objective lens four inches in diameter, is very valuable, but it costs several hundred dollars to buy, while the ordinary amateur cannot hope to make it himself.

But the reflector works on a different principle. It is a shorter, thicker instrument having a large, round, concave mirror in its lower end. The light coming from a star strikes this concave mirror and is reflected upward in a converging cone. Near the upper end of the big tube, which is open at the top, a small diagonal mirror or sometimes a three-sided prism of glass is mounted in such a position that the cone of light reflected by the large mirror is intercepted and is turned at right angles toward the eyepiece in the side of the telescope. Owing to the fact that the light does not pass through the glass as in the other type of telescope, the mirror does not have to be made of optical glass—simply ordinary thick plate glass; and since the mounting of

ON THE SPRUCE-CLAD SUMMIT OF BREEZY MOUNTAIN THE VERMONT ASTRONOMERS HAVE BUILT THEIR STELLAR FANE. A SEVENTY-FIVE FOOT SOLAR TELESCOPE PROJECTS THE SUN'S IMAGE ON A SCREEN INDOORS

In this type of telescope, no tube is required. The light from the sun is reflected by the sixteen-inch, flat, plane mirror B, to an equally large concave, paraboloidal mirror mounted on a stone pier at A. Thence the light converges through a circular opening just above the first mirror, and focuses on the screen, C, where the amateurs study the sun's image

The first page of the November 1925 article in Scientific American

Porter lunching in a restaurant in New York City during this planning stage. To answer some question, Porter got up, took two plates
from the table, inverted one over the other,and proceeded to demonstrate how to grind a mirror. He walked around the table passing the
upper plate over the lower to show the type of strokes he used, much
to the amazement of the New Yorkers around him.

This first article appeared in the November 1925 issue. The response to the story of the Springfield Telescope Makers and their
telescopes at Stellafane was tremendous. So great was it that Ingalls
made a second trip to Vermont to gather more information. At the
November 21 meeting he talked enthusiastically and showed scores
of letters requesting more articles and instructions for making telescopes. They had been pouring in from all parts of the United States
and from around the world. Here was undebatable evidence that the
spark had been kindled, the fire would spread.

Before the November meeting was over, it was decided that Porter would prepare two articles, going into great detail about making
mirrors and mountings. Numerous designs were to be presented for
the telescopes, from the very simple to the complicated, and of
course the articles would be lavishly illustrated with Porter's drawings. He made another trip to New York to discuss the planned
articles, which appeared in the February and March 1926 issues.

In response, letters continued to pour in to the editor's office at
an ever-increasing rate. Alaska, Brazil, Argentina, Australia, Hawaii,
Switzerland, India, England, and Japan were all represented. Although a few amateurs had been making telescopes for years, the
craze was spreading. In anticipation of more articles the *Scientific
American* invited Porter to become a corresponding editor.

It soon became evident that an adequate and reliable source of
materials was needed, from glass banks and grinding abrasives to
mirror-testing advice and instructions on how to construct various
parts of the reflecting telescope. Vice-president John Pierce stepped
in to fill the role and began to ship the various materials across the
country and around the world. At the February 19, 1926, meeting at
Stellafane he reported that he had already filled twenty-three orders,
several from California. He had received sixty-one inquiries from
thirty states and from the West Indies and Hawaii. Others would
soon follow from as far away as India, Tasmania, and South America.

It was not all observing and telescope "shop talk" at Stellafane.

To the club members Stellafane was a place to visit whenever the urge arose. It was a sanctuary of quiet solitude high among the Vermont hills, a place to clear the mind of daily concerns. It was a place of companionship, a place to share a hearty laugh or a hearty supper and a place to share the joys of discovery. Sometimes the members sat by and watched in silent amazement as Porter dashed off one of his countless pencil sketches on any available scrap of paper. The sketch might illustrate an astronomical discussion or a new telescopic idea. Or it might be the consequence of some conversation totally removed from any thoughts astronomical. Sometimes a member had to pose while Porter created a pencil sketch to be mounted near the ceiling of the livingroom with the rest of the portraits.

Judging from club records and other writers, a typical observing meeting at Stellafane might have gone like this: It is late Saturday afternoon, the fall air is cool and clear, and the scattered clouds promise to dissipate after sunset. Most of the members arrive before the evening meal. Everett Redfield, the self-appointed cook and poet, is at work over the wood stove, masterfully preparing another of his varied and savored suppers.

Some of the men are setting up their telescopes on the rocky ledges in front of the clubhouse. Charlie Longe, a machinist, is adjusting the prism in the upper end of his telescope. Porter looks into the eyepiece tube and suggests, "It's still a little misaligned. Tip the prism just a little more in the same direction." The telescope is soon ready for the evening's celestial peregrinations.

To the north and farther down the slope, Oscar Fullam, a pattern maker, has set up his eight-inch telescope. It is a thing of beauty and draws great respect. All the parts were carefully crafted of wood, as only a pattern maker could do, except the metal rings holding the tube onto the mount, the right ascension gears, and the angle bracket needed to place the polar axis parallel to the earth's axis.

John Pierce is standing by and says, "Remember the time, Oscar, at the county fair when someone asked if your telescope was some kind of new butter churn?"

Both men laugh and think back to that day when all the club members had their telescopes on exhibit. Roy Lyon stops by the two men and remarks, "I hope those clouds move out of the way so we can see Venus before it gets dark."

"Yes, and she should give us a fine show today," Pierce adds.

With his white hair, matching mustache, and chef's hat, Everett Redfield appears in the front door. "Supper's ready, come and get it," rings out across the hilltop, and everyone hurries inside. The table is replete with roast pork, mashed potatoes, peas, johnny cake, muffins, and coffee.

Albert Herrick notices Porter adding an excessive amount of pepper to his meal and comments, "Russell, you certainly use a great deal of pepper at supper time."

Porter replies, with a grin, "Well, if you'd spent two winters in the Arctic surviving on nothing but walrus and seal, you'd use a lot of pepper too." Then he reaches for another piece of johnny cake. His favorite section is the crusty corner piece.

Discussions flash up and down the table. Everyday topics having run out, the conversation turns to the latest thoughts on stellar evolution, the aberration of starlight, or the changing length of daylight. John Brashear and other famous men of optics and astronomy watch from their pictures hanging about the room.

Toward the end of the meal Roy Lyon remembers Venus and hurries outside. She is shining brilliantly against the darkened western sky. Hurriedly finishing their dessert, the men gather about the telescopes, taking turns viewing and admiring the crescent of Venus.

Later, Porter points one of the newer telescopes to Vega, no longer overhead as in the summer but moving slowly toward the west in its annual celestial journey. This is the brightest star in the summer sky, and he uses it to check the performance of the telescope by studying its in-and-out-of-focus image. Then while the telescope is pointed in the right direction he looks through the finder and moves it a short distance. He sets the cross hairs between two dim stars at what appears to be empty space, but after a moment of scanning the telescope, he brings into view a tiny smoke ring.

"There it is," he calls out, "the ring nebula in Lyra."

Then, realizing the other men are off at other telescopes, he calls out, "Anyone want to see the ring nebula? It's a beautiful sight in Oscar's telescope," he adds as encouragement. Athough Porter has seen this nebula countless times, he always thrills at its sight.

Another telescope is pointed toward Cygnus, the Swan, or Northern Cross. Albireo at the base of the cross is the target. It is a double star in the telescope, beautiful blue and gold companions.

A voice calls out in the darkness, "I've got M13 in my eight-inch and can resolve individual stars. Anyone care to look?" M13 is the great globular cluster in Hercules, one of the most spectacular sights in a telescope of any size. Countless stars can be seen separately in the outer reaches of the spherical cluster. But the eye is drawn to the crescendo of light at the center, a blur caused by thousands of stars unresolved. To one who has never before seen it, the cluster can produce a strange sense that suddenly some power greater than man is present in the universe, not just in Hercules, but close by in the darkness.

The spell is broken with another call: "I found M81 and M82. Come take a look, Oscar." In the blackness the men cautiously tread the uneven ledges between the telescopes to share their views.

A damp chill touches the late evening air, and Everett Redfield kindles the fire in the fireplace. Soon most of the observers are drawn to it like moths to a light. With chilled fingers surrounding a mug of hot coffee, John Pierce settles into one of the large wooden chairs and says, half joking, "You know, coming in to get warm by the fire is the greatest reward of observing."

Everyone agrees as they stare into the crackling fire. Then Ernest Flanders breaks the silence, "Russell, I never heard how you were rescued from your two-year trip in the Arctic."

"Neither have I," a voice interjects from the dark corner.

Not hearing the second voice, Russell replies, "Well, Ernest, it really isn't much of a story." But after more coaxing he finally launches into the polar adventure. If some of his friends have heard the story before they do not mind hearing it again.

Later someone discovers the old moon rising through the trees. Out into the chill bolt the men, to be repaid by an unusually steady atmosphere and a chance to see the mountains and craters on the eastern face of the moon in sharp relief.

The meeting was planned as an overnight observing session and no one is in a hurry to end the night. But one by one, the men retire to the bunks upstairs. Some wait long enough to get the season's first glimpse of the Pleiades rising in the eastern sky. Soon they will begin to grow dim, but not the memories of this night.

The Garden Telescope, 1921
Porter water color

XI. A GARDEN TELESCOPE AND THE AMATEUR MOVEMENT

Russell Porter was a man always ready to try something new. When increasing years led him to give up demanding global adventure, he turned to the exploration of new worlds of invention and creativity. The most aesthetically pleasing of Porter's inventions is his Garden Telescope. It is also the best example of his skillful blending of art, science, and engineering.

Not long after returning to Springfield, Porter had set about designing a commercial telescope. What he wanted to create was a telescope that would be as ornamental in a garden as it would be utilitarian for observing the surrounding landscape or celestial objects. He felt it should sell for a low enough price so that people of modest means could buy a powerful instrument. At the same time, he envisioned it as a permanent fixture in the garden like the sundial, ready for viewing at a moment's notice. Because it would eliminate the problem of setting up and dismantling a heavy telescope, the Garden Telescope could help to create and perpetuate an active interest in observational astronomy. Hartness liked the idea and agreed to let Porter develop it for production by the Jones and Lamson Machine Company.

In early 1923 the idea passed through the embryonic stage, and the product emerged ready for manufacture. In March of that year an article by Porter appeared in *House Beautiful* describing the new invention, obviously intended to create a potential market of home and garden devotees. In July the *Springfield Reporter* reported that

the telescope was to be made by the machine company in large quantities. "From present indications it seems to be pretty well settled
that the manufacture of the 'garden telescope,' . . . is far past the
toddling stage." Full of optimism, the story went on to say that "it
bids to become one of the leading industries of Springfield."[1] Unfortunately, it did not live long enough to fulfill such grand prophe-
cies.

So different from the ordinary telescope is the Garden 'Scope that
on first sight it might not be recognized as what it is. Gone are the
lenses and long tube of the normal refracting-type telescope. Gone is
the tube of the Newtonian telescope, even though it is of this type.
The six-inch aperture mirror is the heart of the telescope. A bowl of
bronze lotus leaves embraces the mirror so that it pivots about itself
when the telescope it pointed to different parts of the sky or landscape. Reaching out from the side of the lotus bowl, a slender bronze
leaf curves gracefully upward about two feet to a prism and the eyepiece. Different eyepieces allow one to change the magnifying power
of the telescope between 25, 50, and 100 times. This is adequate to
show such celestial marvels as the mountains and craters of the
moon, the satellites of Jupiter, the rings of Saturn, star clusters, and
galaxies.

If the telescope is pointed at the sun and the image projected
onto a white paper, the instrument becomes a sundial, and the sun
time may be read from the hour circle that encircles the lotus bowl.
A table of numbers provided with the instrument allows the sundial-
ist to obtain standard time. Roman numerals show the proper hour,
and graduations divide the circle into ten-minute intevals.

The hour circle, which provides the right ascension of the telescope, has a segment removed so the bronze leaf holding the eyepiece can pass through. This allows the telescope to point to all parts
of the sky and landscape. Thus Porter made it possible to have the
center of mass of all the moving parts within the mounting points of
the instrument, a necessary feature to provide a sturdy, compact, and
vibration-free instrument. Actually it was an application of the design he had published in *Popular Astronomy* and *Scientific American*
in 1918 for a new mount for large telescopes. The simplest ideas are
always the best, and it was this scheme that Porter later proposed for
the mounting of the 200-inch telescope to be constructed in California.

Detailed bronze leaves and a massive geometrical base disguise the fact that the mounting is actually a combination of the alt-azimuth and equatorial design. For daytime viewing about the landscape the alt-azimuth form is used. Because one of the two rotating axes is then vertical, the telescope is easily scanned from object to object. When night descends and the celestial sights are to be observed, the vertical axis is clamped so the telescope acquires the equatorial mount. Now it is necessary to turn only the polar axis to follow the stars as they file across the sky. This is done by manually turning a knob in the shape of a bonze flower on either of the bearings supporting the hour circle.

The third axis, the declination axis, is accompanied with its setting circle, which passes half way around the lotus leaf bowl. Using this circle and the hour circle, and knowing the siderial time, the viewer can set the telescope on any celestial object and it will be in the field of the low-power eyepiece. Hidden from view is an adjustment for initially setting the telescope to any latitude between twenty-three and fifty-five degrees. All parts are cast in statuary bronze so the telescope can withstand any weather without maintenance. The names of Kepler, Newton, and Galileo, three of the men who did most to shape the thought of modern astronomy, are cast into the base of the telescope, symbols of the solid foundations they created in astronomical thought.

Henry Loudon, who was then in charge of advertising for the Jones and Lamson Machine Company, recalled years later that there was little or no interest in the Garden Telescope within the company; only Porter and Hartness were enthusiastic. Others felt it had little chance of becoming a big selling item and did not fit into the company's sales and promotional activities. Nevertheless, the telescope was put into production.

The optical shop, recently created to produce the comparator, was put to use in the fabrication of the mirrors. A young lad, Wilbur H. Perry, who had joined Porter's group of amateur telescope makers, was responsible for parabolizing all the telescope mirrors. (He later went on to become chief technician of the Ruling Engine Laboratory at the Johns Hopkins University.) The company was not equipped to make the prisms or eyepieces, so these were contracted to John Brashear, the famous American telescope maker from whom

Porter examining the sun with an early model of his Garden Telescope, circa 1921

Porter with his Garden Telescope, circa 1923

Hartness had earlier obtained the sixteen-inch mirror blanks that Porter used in Port Clyde.

Once the telescopes were sold, their owners were never heard from again, and no one knew whether they were actually being used. It was finally decided that Henry Loudon should take a trip across the country to learn what became of the instruments, with the hope of generating more sales. He traveled through the South as far as New Orleans and continued on to the West Coast, visiting the people who bought the telescopes to find out if they knew how to operate them—all the while remaining a salesman. But new sales were barely enough to cover his travel expenses.

The Porter Garden Telescope as it appears today

What he discovered was that the people did not understand the operation of the telescope, particularly the equatorial mounting. Some were being used only as garden ornaments, while others were still in their boxes. He returned to Springfield and with Porter rewrote the instructions, enabling future customers to set up and use the telescope properly.

Because the company records have been lost, there is no record of the number of Garden Telescopes that were manufactured or of when the product was discontinued. Estimates range from 75 to over 200 sold at as much as $500.00 apiece. The highest remaining serial number is fifty-three, so at least that many were made. In spite of Porter's original hope, the price was more than the average telescope user could pay and may well have been the cause of the poor sales. Today there are very few Garden Telescopes still in existence; the known and suspected ones can be counted on two hands. It is a sad end for an ingenious and lovely idea.

Other Porter projects were flourishing, however. By the summer of 1926 Albert Ingalls and the *Scientific American* had uncovered the large number of hobbyists making telescopes across the country. Porter's fellow amateur telescope makers decided it would be fun to invite them to Stellafane for a weekend dedicated to the exchange of ideas and experiences. The date was set for July 3, 1926, and the enthusiasts drove into Springfield and up Breezy Hill, following the dirt road through open fields until they reached the broad top of the hill and Stellafane. Some even brought their telescopes and drove very cautiously to protect their delicate offspring from the bumps and ruts.

That day they compared telescopes, talked of mirror making, camped, ate, talked some more, observed the planets through black sky, and completely forgot the world in the valley below. By later standards that first convention was a tiny gathering, twenty registered persons. Most were from New Engand and New York, but one came from as far away as Virginia. Nevertheless, the gathering was the highlight of the year for the Springfield Telescope Makers. They thoroughly enjoyed themselves, acting as hosts and sharing in the exchange of ideas and methods where few instructions existed.

At the end of the summer the club resumed its regular monthly meetings and special observing sessions at Stellafane. Sometimes they worked to improve their clubhouse, adding, for example, a workroom

The first gathering of telescope makers at Stellafane on July 3, 1926
Porter seated in front, and Ingalls, wearing glasses, seated at center

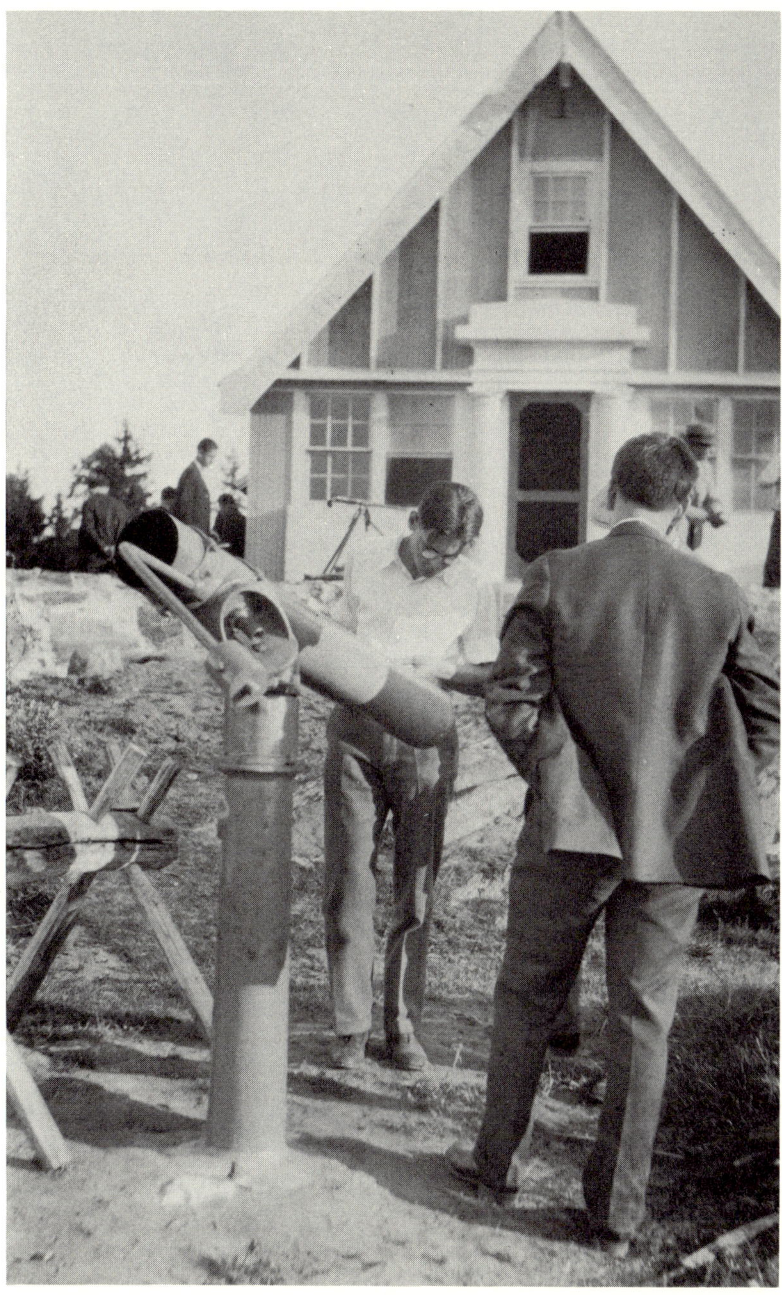

The Springfield telescope, invented by Porter, being examined at the first convention of amateur telescope makers at Stellafane on July 3, 1926

Porter, right, talking "telescopics," at Stellafane, July 3, 1926

*Ingalls, left, and Porter posing proudly with Porter's invention, the
Springfield telescope, at Stellafane
  Date unknown*

off the kitchen. In March 1927 it was voted to accept Albert Ingalls into the Springfield Telescope Makers as an honorary member.

At the same meeting Porter unrolled some full-scale drawings he had obtained from Dr. George E. Hale of Mount Wilson Observatory, who at the time was trying to interest people in his plan to build a 200-inch telescope. The drawings showed Hale's design for a spectroheliograph, an instrument for photographing the entire surface of the sun at any one wavelength of light. Porter was fascinated by the instrument's capabilities and eager to construct the new astronomical tool. Although work was done on the instrument by club members over several years, it was never completed. The diffraction grating, which had been a gift from Dr. Hale to Porter, was destroyed by fire while on loan to an amateur solar observer.

Porter and his friends continued to extend their interest in telescope making beyond the Green Mountain State. During the spring of 1927 they received word from California that another amateur club had been formed at Riverside. This was one of the first clubs founded for the purpose of making telescopes after Stellafane was created. Their hobby had spanned the entire continent in four years.

On Friday, April 15, 1927, Porter and several other Springfield men journeyed by bus to Cambridge, Massachusetts, to be guests of the Bond Astronomical Club of Harvard Observatory. That organization of amateurs had been formed by the world-renowned astronomer and director of the Harvard Observatory, Dr. Harlow Shapley. Porter gave a talk about the history of Stellafane and Vice-president Pierce talked about mirror making. Also present were James Hartness, who talked about his underground observatory, Harlow Shapley, and Albert Ingalls. Astronomy is one scientific field where the professionals take a sincere interest in the amateurs. Beyond the pleasure of sharing knowledge with others, they do not waste their time. From William Parsons, Earl of Rosse, to Clyde Tombough, discoverer of Pluto, the history of astronomy is studded with men who have left the ranks of amateurs to join the professionals.

During their visit to the Boston area, the Springfield guests saw pioneer work being done with fused quartz at the Thomson Research Laboratory of the General Electric Company of Lynn. Quartz was known to possess the desirable characteristic of changing its size by only a negligible amount when its temperature changes. Thus it does not break when faced with a sudden shock from heat and for this reason had found its way into scientific laboratory apparatus. This characteristic also made fused quartz the ideal support for the silver reflective layer of the astronomical telescope. Unlike the plate glass and Pyrex glass then being used, a quartz mirror would not warp or refuse to hold the reflecting silver in the precise parabolic shape that its creator had labored so long to produce. This would be particularly valuable in large reflecting telescopes where a change in the evening's temperature can strain and twist the mirror until the entire instrument becomes useless.

Professor Elihu Thomson, inventor of electric welding and numerous electrical devices, was well acquainted with the virtues of quartz. As an amateur astronomer he had built his own observatory, and for the past twenty years he had been experimenting with fused quartz. Now at General Electric he was trying to make discs of the material large enough to be useful in telescopes, but it was difficult to fuse into large pieces. Quartz, or silicon dioxide, contains the two most abundant elements on earth and yet it continued to resist man's struggle to shape it into useful optical components of large size.

Of course that visit to the General Electric plant was made before the amateurs knew much of Hale's plans for the world's largest telescope. But soon the designers would be asking if their mirrors could be made of fused quartz. It is now well-known that Elihu Thomson was confident the 200-inch disc could be made of fused quartz at his plant and that a great deal of money was spent in the effort, but in the end the mirror had to be made of Pyrex. Quartz would continue to resist man's technological probing for years.

In June 1927, Professor Leon Campbell of the Harvard Observatory accepted an invitation to visit Stellafane. He gave a lecture on observing variable stars, hoping that some of the telescope makers of Springfield might be motivated to contribute to this valuable study to which amateur astronomers were making significant contributions by recording the periodic light fluctuations of these stars. But there is no record that the Stellafane group ever allowed their interest in instrumental tinkering to be threatened by such a discipline.

On the weekend of July 9–10, 1927, most devoted amateurs in the East again ascended to Stellafane for the second convention of telescope makers. The number of registered persons soared to fifty-nine, and even more were present, counting families. Minnesota was the most distant state represented. Those who were returning looked forward to meeting old friends and swapping the year's stories about telescope making. The newcomers were eager to glean what they could from those with more experience. The hosts provided good food, good camping, and good observing. Most important of all, they provided the chance for everyone to learn by talking with everyone else and have fun doing it.

Porter was actively present, always willing to listen to anyone's troubles and lend advice and encouragement where needed. Of all the amateurs present, he was certainly the most experienced in making and designing telescopes. People came to him for help in dispersing the fog of doubt about how to "read" the light shadows in the delicate mirror test, or to be advised on how to get a full polish on a mirror. Or Porter might dash off a sketch to show the importance of making a sturdy mount. He was interested in each telescope brought to the convention and how it had been created by its maker, always eager to demonstrate the joys of accomplishment and to inspire people to continue their efforts.

The second convention was such a success that everyone wanted to make it an annual affair. They were already looking forward to the following year when they could again return to Breezy Hill and Stellafane. This affection for the Stellafane conventions has continued through the years and has grown until today hundreds of amateurs make the annual pilgrimage from all parts of the country and Canada, eager to show, talk, share, learn, observe, and literally eat and sleep telescopes with companion TNs (Telescope Nuts).

When the Springfield club members were not hosting the gathering on the hill or building and using individual telescopes, they were designing and fabricating instruments as club projects. Here was another way to increase the enjoyment of their hobby. With many hands to help they could build larger telescopes and with the pooling of resources they could exchange ideas. At various times club telescopes, all of different design, were assembled and tested at Stellafane. For example, at the upstairs window on the south side of the clubhouse they mounted a sixteen-inch flat mirror as a coelostat to direct sunlight in a sixteen-inch paraboloid mounted on a concrete pier seventy feet to the south of the building. The paraboloid of seventy-five-foot focal length returned the sunlight to the upstairs window, where an image of the sun over seven and one-half inches in diameter was formed in the darkened room. The sun's image could be observed by several people at one time. For best resolution the mirror had to be stopped down to a four-inch aperture.

During his efforts to improve the convenience of observing, Porter designed a polar-Cassegrain telescope. This would provide a fixed eyepiece and high magnifying power in a compact instrument. The members set to work building the parts. Finally they erected the instrument so it protruded from the south of the clubhouse through an opening in the kitchen wall. A sixteen-inch flat sent starlight up the polar axis to a perforated twelve-inch paraboloid of forty-eight inch focal length. This reflected the light to the hyperboloidal secondary, which in turn directed the light back up the polar axis to a focus at the eyepiece in the kitchen. For convenience they installed slow motion controls so the telescope could be run from the viewing location inside the building.

Porter's teachings in telescope making were beginning to spread around the globe as a result of his articles printed in *Scientific American* in 1926. In addition, thousands of amateur telescope makers within reach of the periodical were fortunate to have in Albert Ingalls one of the most active hobbyists of all. He had struggled with his own telescope mirror when instructions were practically non-existent in the United States; there were only the two articles by Porter in *Popular Astronomy*. Now Ingalls was eager to help Porter communicate with the worldwide fraternity and was in a position to do so through the pages of *Scientific American.*

The response to those first two *Scientific American* articles by Porter had been so rewarding that Ingalls edited a book called, appropriately, *Amateur Telescope Making*, published by the Scientific American Publishing Company in 1926. Included in the little book were Porter's articles and several new chapters. Porter wrote about how easy it was to make a 100-foot focal length sun telescope capable of producing an image of the sun over ten inches in diameter. The scheme had been used successfully at Stellafane. He told the amateur how to make flats for his diagonals and lenses for his eyepieces. He described some of the problems of making a Cassegrain telescope, of which he had made three, and he did not hesitate to point out that they never performed as well as an equivalent Newtonian telescope. The 102-page book, of which only 3,400 copies were printed, also contained sections of the Reverend William F. A.

Ellison's book, *The Amateur's Telescope*, published in London in 1920. Ingalls' book was sold out by 1928.

The sale of the book was proof that something was happening. It was Porter's arctic fever all over again, only more contagious. During winter observing sessions, there are in fact other similarities to the Arctic as the enthusiastic amateur spends his share of nights out among the stars. Each marvel of the universe serves to stir the imagination further and beckons the explorer to seek behind the ranges.

By May 1928 Ingalls had founded a regular department in *Scientific American,* devoted entirely to amateur telescope making. Now at last the amateur could turn monthly to his mail and "talk shop" with thousands of other enthusiasts. He could learn from the experiences of others and study pictures of telescopes made by other TNs. He could have the feeling that he was sharing with others the wonderful world of scientific exploration.

Porter continued as editor to the magazine and contributed greatly to the amateurs' department. He offered countless sketches to help clarify points in mirror making or to show ideas for mounting. Occasionally he would even pass on one of his cartoons depicting the backyard adventures of observing amateurs.

The highlight of the summer of 1928 was the third annual convention of telescope makers at Stellafane on July 7. But even then events were slowly transpiring that would alter Porter's life more than he had ever dreamed. At fifty-six, he was about to start the last great adventure of his life. He was about to join the team of scientists and engineers being formed to create the world's largest telescope.

XII. CALL TO CALIFORNIA

Before Russell Porter and George E. Hale actually met, they were acquainted indirectly. It was Hale to whom Porter had written in 1927 requesting the plans for his spectroheliograph. Not only had Hale seen Porter's articles on designing telescopes in *Amateur Telescope Making*, he was also an editor of *The Astrophysical Journal* when Porter's article on photographing the knife-edge shadows appeared. Then, in early 1928, Albert Ingalls set up a dinner for the three of them in New York City. Ingalls later wrote in *Scientific American* to all his TN friends that during the meal Porter had talked impressively about his ideas for large telescopes, all the while illustrating them with quick sketches on every available piece of paper including the menu.

In the fall of 1928, Hale sent two of his scientists, John A. Anderson and Francis G. Pease, to Springfield. (The story has already been well told by David O. Woodbury in his book *The Glass Giant of Palomar*.) Coming to visit Porter and discovering that he had gone on a picnic with his family, the two men were directed to Hartness instead. He was told the nature of the astronomers' visit and did his best to "sell" his man. When Porter returned and was sent for to meet his guests, Hartness intentionally did not bother to say who they were. There followed a congenial visit. Porter talked freely about everything from Vermont to the Arctic to being an artist, but little, if anything was said about astronomy, let alone telescopes. Not until the visitors were on the train did Hartness divulge

Telescope designed by Porter for site survey for the Hale telescope

that they might be interested in having Porter join the staff being organized to build a 200-inch telescope. Porter was speechless. What could he contribute to such a vast undertaking? He was just an amateur and they were professionals.

Many weeks went by before Porter heard from the astronomers of the California Institute of Technology. At last he received a telegram from Hale dated November 17, 1928. It read: "Can you come to Pasadena for several months to assist in designing two hundred inch telescope and instrument shop auxiliary instruments. Please wire collect monthly salary desired and whether you can come at once as we wish to push work rapidly." [1]

An exchange of messages negotiated salary, and then Porter received a telegram dated November 21. In part it read, "Please come at once to stay until July first. Salary six hundred per month plus your own personal travelling expenses both ways." Hale closed with "Advise do not bring household goods but rent furnished house." On November 23 Porter sent a final telegram to Hale: "Terms of your nightletter accepted stop We leave here Monday and are due to arrive at Pasadena via Union Pacific at ten twenty Saturday morning December first." [2] His good friend and fellow telescope maker Oscar Marshall once quoted Porter as saying he did not hesitate to undertake the task. This he had said with not the slightest show of emotion or pride during a farewell dinner given by the Springfield Telescope Makers.

Porter once wrote that he had studied architecture so he could design forty-room villas but ended up designing three-room cottages. He went from the sublime to the ridiculous. Then he started making small telescopes and now was about to start work on the 200-inch; now he was going from the ridiculous to the sublime.

Porter did not make the move to California without feeling some regret at leaving his friends at Stellafane, men with whom he had shared many happy hours. But he was not leaving them entirely on their own. He planned to keep in touch and to return the following summer. Vice-president Pierce assumed the duties of president in Porter's absence.

No sooner had Porter and his family made their hasty move than Hale put him to work. His first task was to design a small telescope to be used for "seeing tests," which would help determine the final site for the 200-inch telescope. This was a four-inch refractor with a magnification of 750 diameters. A telescope with such a large magnification allows an observer to study the diffraction rings about a star image and to measure image motion due to the turbulent atmosphere. In this way he can assess the ability of the atmosphere to pass steady and sharp images at any desired site. As a matter of convenience Porter designed the refractor to look only at Polaris. Ten or twelve telescopes were finally made and used for testing sites throughout the southwestern United States in 1929 and 1930. Porter also designed two twelve-inch Cassegrain telescopes for site testing. Later, when the testing had been completed, one of these Cassegrain telescopes was converted for use as a finder on the 100-inch Hooker telescope on Mount Wilson.

It seems unnecessary to say that after the tests the choice was Palomar Mountain, southeast of Pasadena and northeast of San Diego. Far enough away from city lights to guarantee dark skies during any foreseeable future, yet close enough to an astronomically oriented educational institution to be readily available to astronomers, with a large plateau top and excellent seeing throughout the year, the mountain was an ideal location for the world's largest astronomical observatory. The geologists said it was a huge granite block between two faults, a geological island in an earthquake-infested sea; nothing better could be found in Southern California.

Dr. Anderson, one of the men who had visited Springfield, became Porter's immediate boss. The title of Associate in Optics and Instrument Design was given to Porter.

The 200-inch project was going to require more space than was available on the Cal Tech campus. This meant more buildings were going to have to be designed and built. Space was needed for staff headquarters, offices, laboratories, model-building shops, wood and metal working shops, and of course a large optical and testing facility. Three buildings were decided upon, and Porter set about making the preliminary sketches. He took great care to see that they would be modern yet blend in with the surrounding campus buildings. His plans were then turned over to Goodhue Associates, an architectural firm of New York City, which produced the final designs.

Porter completed the plans for the astrophysics laboratory first, then for the machine shop and the optical shop, incorporating ideas of several members of the staff. Within the windowless confines of the latter, all the optical elements for the 200-inch telescope would be ground, polished, figured, and tested. Very special problems had to be faced and overcome in the design of this shop. The 200-inch mirror would be tested with the Foucault knife-edge test, just as amateurs and professionals had been doing for years. For the tests to be precise and meaningful, the path that the light would take to and from the mirror, about 225 feet, had to be absolutely homogeneous. This required an accurate control of temperature and humidity, which was partially realized by leaving out the conventional windows, lining the interior of the concrete walls and ceiling with a three-inch layer of cork board, and pouring the concrete floor on a special insulating sandwich of cork board, sand, gravel, and crushed rock. The control was completed by giving the building its own heating, cooling, and air conditioning systems so the 53'×165'×39'-high polishing and testing room for the 200-inch mirror could maintain the desired conditions.

The machine shop, a one-story building 50'×200', would be completed and working by October 1930. Later that year the five-story astrophysical laboratory would be begun, and by early 1932 it would be completed. The optical shop would not be completed until the summer of 1933.

During this period Porter occasionally did some astronomical observing. For example, in early July 1929 he went up to Mount Wilson to use the sixty-inch reflecting telescope to make lunar drawings. Some of these drawings were later included in an article by him in *Scientific American* for October 1930. "Previous work of this kind, done some years before, together with a facility in using a pencil, made it seem likely," Porter wrote, "that provided the seeing was good, I would be able to transfer to paper certain lunar objects or markings which would be of value and serve as a kind of standard of seeing." However, when his alloted time came to use the telescope, the seeing was poor and the sketches could be of no scientific value. But Porter still delighted in making "interesting sketches." During these early morning hours spent at the eyepiece, he could not help but relate the scene to his own earlier experiences. "These recent nights with the moon vividly recalled the trackless wastes of arctic snow and ice, because of the similarity of the desolate and dead moonscape to our arctic regions." [3]

Porter had been hired originally to work only until July 1, 1929.

Then he was free to return to Springfield, Vermont, for what remained of the summer. He would have to wait for word from Hale concerning the Observatory Council's decision to rehire him for the coming academic year. In the meantime he spent his time working for Hartness and going to New York City to discuss design problems with the architects for the Cal Tech buildings. Toward the end of August he received a letter from Hale stating that he was being offered a position for three years at $5,000 a year for the academic year. Porter readily accepted.

While working on the plans for the three buildings, he had been thinking up schemes for the mounting for the 200-inch telescope, although from the very beginning he had proposed his split-ring mount, the same principle used years earlier in his Garden Telescope. Now, while he was back in Springfield he talked about his designs to Hartness, who was keenly interested in the entire exciting undertaking. Hartness felt that one of Porter's plans had so much promise that he sent him into the pattern room to make a model.

Before returning to California, Porter, accompanied by Hartness, visited the General Electric Company in Lynn, Massachusetts, to check on their progress with quartz mirror blanks. He was able to report to Hale that Elihu Thomson and his fellow G.E. workers had made half a dozen twenty-two inch discs of various degrees of excellence. They were also planning a furnace to handle a sixty-inch disc. Thomson was very optimistic of their ultimate success with a 200-inch disc.

Back in Pasadena, Porter continued his architectural designs for the three buildings and his mechanical designs for the 200-inch telescope mount. These took most of his time during that academic year. The next summer in Vermont, 1930, was mere repetition. But on his trip back to Pasadena, Porter acted as Hale's emissary, stopping off in various cities to discuss with experts the problems of building a grinding and polishing machine large enough to handle a 200-inch mirror. Among his stops were Pittsburgh and the J. W. Fecker Company, and Cleveland and the Warner and Swasey Company.

Nothing much is known about Porter's contribution to the 200-inch project during the academic year 1930–1931. But during the next summer visit to Vermont, he again took the time to go to the General Electric Company and inspect the most recent quartz mirror blanks. Unfortunately he was not able to give Hale an encouraging report. Thomson and his men had worked their way up to the sixty-inch diameter blank needed for the 200-inch telescope, but the stubborn quartz had twice refused to cooperate. Both blanks had produced a crack. The second had suffered from brick particles falling from the roof of the furnace while the mirror was being made. Porter was able to report only that he thought the blank would make a serviceable mirror. But the blanks were never used by the Cal Tech group, and they lay in the G.E. plant for years. The 200-inch quartz mirror was still a long way off.

This is the story of Porter, not the 200-inch telescope, and of necessity many of the fascinating details of that story must be left out.[4] Suffice it to say that hopes for a quartz mirror were finally abandoned by the end of 1931 after much money had been spent. The Corning Glass Company was chosen to do the task. This time Pyrex glass, less sensitive to temperature changes than plate glass but not as good as quartz, was agreed upon. After two tries, Corning was successful, and the 200-inch was born.

In 1932 a young physics research fellow at Cal Tech came up with an idea that was to become a major contribution to the whole field of optics. Dr. John Strong felt he could coat the mirrors for the reflecting telescope by placing them in a vacuum and heating silver until it boiled and evaporated, spreading evenly throughout the vacuum and thus coating the mirrors with a thin silver reflecting surface. To test out his ideas Strong needed a vacuum system and bell jar. Porter was asked to design the first bell jar. It was an eighteen-inch diameter affair, and when he finished the design he went to the shop and made the patterns for the castings.

Porter always reveled in the chance to work with his hands whether with glass, metal, or wood. He even had his own space in the pattern room where he could work up patterns or make wooden models. In early 1932 he wrote to his friends in Springfield, Vermont: ". . . and thanks for shipping the tool box. Came O.K. and now it's on my bench in the pattern room ready for work." Some of the workmen made "a fine pattern bench . . . vice and all. Guess I bother em so bad they had to get rid of me." He went on to say how much he had been involved in model making. "My office is just filled with small models I've made from time to time of machines, mountings, telescopes, etc."[5]

Porter working in his office at Cal Tech, Pasadena, 1935

Porter studies for a solar telescope, 1932

Coelostat designed by Porter at Cal Tech

After Strong showed that his evaporation process was able to produce a better coating with less work and under less dangerous conditions than the old chemical deposition method, he went on in the following year to make mirror coatings in aluminum. These proved to have the higher reflection in the utraviolet and to be longer lasting than the old silver coatings. The results were so successful that all the mirrors for the 200-inch telescope were to be coated with aluminum.

Porter's three-year employment agreement with Cal Tech terminated by the summer of 1932. With a depression on, uncertainty hung heavy over the Porters as they drove east for the summer. When they arrived in Springfield, Russell discovered Hartness could not employ him during the summer at the Jones and Lamson Machine Company because of the Depression. At Port Clyde, however, he received a letter from Hale dated July 28, 1932, stating that his work agreement had been renewed for the coming year. At Porter's request, Hale told him to plan to be in Pasadena for the following summer also, even though he could not give Porter a salary for those months.

All through the period, countless designs for the mounting of the 200-inch telescope were being made, discussed, and altered, and Porter was involved in many of them. Oscar Marshall, who had joined the team in October 1930, at Porter's recommendation, wrote proudly in late 1932 of Porter's increased value to the project. "Something different in the way of a mount for the 200″ is of course brewing much of the time, and you all know that for such a dilemma, Russell always has a variety of answers." [6]

Marshall was a valuable member of Hale's project. He brought with him forty years of practical experience as a journeyman machinist, the major part of the time spent in charge of machinists and writing articles for trade journals. He was selected to machine most of the thousands of precision parts required for the thirty-six mounts on the 200-inch mirror. In addition, he did all the gear cutting that could be done in the machine shop. When the capacities of the machines were strained, he designed attachments to handle the larger gears. Like Porter, he also was an artist who loved nature and the wilderness. He was an oil painter, photographer, and writer interested in the history of optics. At one time he was the official photographer for the 200-inch telescope.

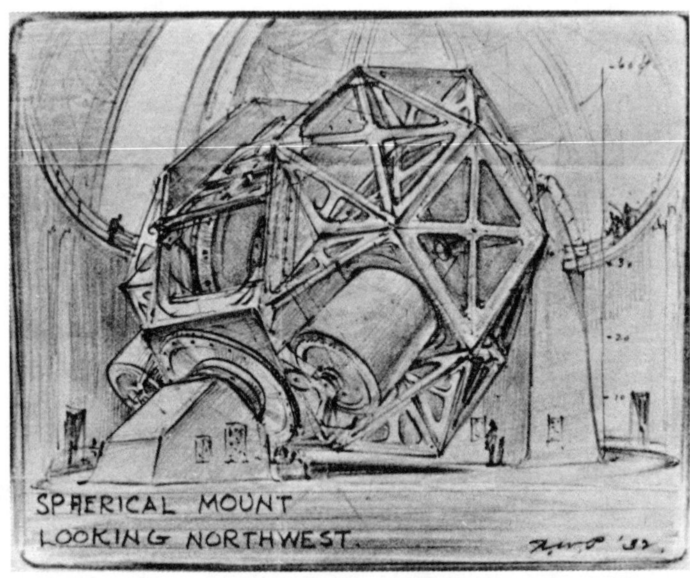

Rejected Porter design and model for spherical mount

A project that occupied Porter during the winter of 1934 and finally yielded to his determination was the fabrication of a small aspheric corrector lens for the 100-inch Hooker telescope on Mount Wilson. In July 1934 he wrote to his friends at the annual convention at Stellafane: "I have a distinct recollection of being incarcerated for two months the past winter in the basement of the optical shop working on one surface of a two inch lens." [7] When the lens was finally finished it was used in the photography of the spectra of distant nedulae so their red shifts, and hence their distances, could be determined.

In January 1935 Porter wrote to the secretary of the Springfield Telescope Makers that he had become involved with a new type of telescope that would supplement the work of the 200-inch. Actually, this new instrument was a wide-field camera and not a telescope and had been invented recently in Germany by Bernhard Schmidt. Now Porter was designing an astronomical camera based on this new principal. "It is a combined reflector and refractor, there being a very thin negative lens at the outer end of the tube where we usually have our eyepiece." He went on to explain, "The mirror is spherical and the lens adds to the mirror just what parabolizing does—with this exception, that the field of view for good definition is greatly enlarged for photographing." He explained how the comet-shaped star images formed by the normal reflector are corrected to sharp points of light in the Schmidt camera and how this is accomplished by placing the corrector lens at the center of curvature of the mirror. This scheme was so uniquely simple in the field of optical design that it was overlooked for many years. "The difficult job—optically—is figuring the thin lens which is only a quarter of an inch thick. The scope we are to build has a 26″ mirror, 18″ correcting lens 3 ft. focus." Porter was given the chance to do the optical figuring and was eager to get back to "pushing glass." "They are building me a machine now and, when ready, I'll get into my frock and try my hand at optical surfaces." [8]

Later in the year Porter took time out from his work on the mounting designs to visit Palomar Mountain. With the assigned task of surveying the uninhabited plateau he took his surveying transit and plane table and camped alone in a tent until the job was completed. Working outward from the site chosen for the 200-inch telescope he laid out the locations for the auxiliary equipment and

An early Porter study of the observers' cage and elevator for the 200-inch telescope

Porter drawing of the observers' cage and elevator for the Hale telescope roads. He chose the sites for the eighteen-inch Schmidt camera and the forty-eight-inch Schmidt camera that would one day serve as scouts for the explorations of the 200-inch telescope. He laid out the water tower, machine shops, garages, and other service buildings.

In a pine woods Porter set the living quarters for the staff and the visiting astronomers. The visitors would stay in the "Monastery," so named for its bachelor quarters. The staff and their families would live in five cottages spread out among the trees, with plenty of space between, somewhat reminiscent of the cottages at Port Clyde.

After returning to Pasadena, Porter made a contour map from his survey and a landscape drawing. Then he made a scale model of the whole mountain top, showing exactly where each building would be located. Although he did the complete architectural plans for the housing he left the details to someone else. For the Monastery he urged the Observatory Council to build the most modern building available and won. He went ahead and designed the house complete with eight bedrooms, living room, library, recreation room, electric kitchen, and quarters for a couple to run the place. At his insistence he was allowed to spare nothing in providing for the complete comfort and relaxation of the visiting astronomers.

The design for the mounting of the large telescope had been considered, of course, from the very beginning. In 1928 the choice had already been narrowed down to two, the fork mount like the sixty-inch telescope on Mount Wilson, or the yoke mount like the 100-inch telescope on Mount Wilson. Each had its own advantages and disadvantages. In the fork mount the tube of the telescope is held by two outstretched arms reaching for the north celestial pole. Since all of the mounting is south of the tube, most of the observatory floor is left free for working with accessory equipment. But the center of mass of the telescope is not between the bearing points. This creates an unstable instrument. With a telescope as large as 200 inches it might not be possible to prevent the mount and tube from flexing out of alignment as it was moved about the sky.

In the second mounting, on the other hand, the fork is extended north and closed to form a yoke, which in turn is held in bearings at both the north and south end of the polar axis. Now, the center of mass is held between the bearings and remains in a stable position. The sacrifices for rigidity are that the telescope cannot be pointed near the north celestial pole since the tube would collide

Cutaway drawing by Porter of the 200-inch Hale telescope, 1938
  Hale Observatories photograph

Cutaway drawing by Porter of 200-inch telescope mirror and cell, 1937
Hale Observatories photograph

with the north bearing, more of the observatory floor is used up for mounting, and less is available for working.

After the two mounting possibilities were selected, much design effort was put toward the fork concept. For a while the Warner & Swasey Company, who had many years of experience designing and building telescope mountings, was called in for consultation. Their assignment was to come up with a fork mount, but since they refused to guarantee the rigidity of such an approach their contract was cancelled.

Even while expending much effort in favor of a fork mounting, the scientists and engineers at Cal Tech could not forget the yoke mount, and there was some vacillating between the two. During this time Porter was kept busy making sketches and drawings of all the various schemes that were thought up. Most valuable perhaps were his three-dimensional cutaway drawings, which showed exactly how a particular idea was to work. When he was not making drawings he would go down to the model shop and turn out wooden or metal models.

Of the men who had been closest to the conceptual problems of the mounting, Porter and Francis Pease, who had interviewed Porter in Springfield, had mulled over the challenge the longest. Independently, they had worked out similar mounting schemes. In fact, so nearly identical were the designs that it is impossible not to wonder whether each knew of the other's ideas.

Over ten years before Porter had gone to Pasadena, in March 1918, he had published his article in *Popular Astronomy:* "A New Form of Mounting for Large Reflectors." It had described how in the new mounting he placed the tube within a large ring that was split to allow the tube to pivot north and south. The split ring acted as the upper end of the polar axis and a thrust bearing at the lower end. It rotated on two ball bearings situated about its perimeter. This type of mounting was well suited for large telescopes, as Porter pointed out, because it was compact, the center of mass was between the three bearing points, and the tube could be pointed to all parts of the sky including the north pole. Soon after the article appeared, this principle was put to a small-scale test in the Garden Telescope.

Dr. Pease had also been thinking about mountings for large telescopes before he began work on the 200-inch telescope. According to a sketch dated November 7, 1921, he had already worked out a pre-

A typical Porter pose inside the 200-inch telescope dome, circa 1939

liminary design for a 300-inch telescope. Although the mounting was very close to that of Porter's little Garden Telescope, Pease could not have seen Porter's patent for that because it was not filed until January 25, 1922, and awarded September 25, 1923. Of course Pease may have seen Porter's 1918 *Popular Astronomy* article. But if there ever was any exchange between these two men before 1928 concerning their telescope designs, no record has yet been found.

The whole process of developing the final mounting is best described as evolutionary. Neither idea was finally accepted but rather one that could claim an ancestor in both the fork and the yoke schemes. Imagine the upper bearing of the fork mount enlarged to a ring larger than the telescope tube and its upper quarter section removed to form a split ring. The ring rolls on two bearings. Imagine the fork arms shortened and evolved to extinction. Now the telescope tube pivots within the split ring and within the bearing points, as in Porter's Garden Telescope. Or imagine the upper bearing of

Porter testing a mockup of the 200-inch telescope prime focus at the Cal Tech machine shop

Porter drawing of the 200-inch telescope prime focus, 1940
Hale Observatories photograph

the yoke mount enlarged greatly and its upper quarter section removed so the bearing surface is now the outer edge of the ring. Again the tube pivots within the three bearing points and can be pointed to the north celestial pole, a modification of Porter's Garden Telescope. So each approach can yield the same result, which came to be called the Horseshoe Mount because the split ring took on the shape of a "U" or horseshoe. As in most team projects no one person could claim parentage of the final Horseshoe Mount.

Early in 1936 the Westinghouse Electrical and Manufacturing Company of Philadelphia began construction of the parts for the mounting. By August 1938 they were completed, and in the following November they arrived on Palomar Mountain. Even before the designs for the mounting were finished in late 1935, the design for the dome of the observatory for the 200-inch telescope was completed. Like the optical shop, the dome had to do a superb job of insulating its contents, for the telescope could not be allowed to heat up during the day or it would be useless during much of the night. The dome was designed essentially as a double shell with air circulating between to remove the heat absorbed from the sun. Although the entire engineering staff contributed to the design of the dome and observatory, Porter was responsible for many of its details. As with the telescope mounting, one of his most important donations to the project was his cutaway drawings, which seemed to be always present whenever a dispute or misunderstanding arose.

For the main entrance to the observatory Porter designed a small and simple portal, almost insignificant underneath the huge dome that towered 135 feet above. Around the perimeter of the stationary part of the circular building he placed the photographic darkrooms, kitchenette for midnight snacks, and other rooms necessary to house the various accessory controls and equipment. Wanting to keep the curious and interested public informed yet out of the way of astronomers and away from the thermally delicate telescope, Porter designed a glassed-in observation deck reached by a special stairway. Here visitors could go no farther but could clearly use the huge telescope towering high above them poised for the evening's journey into space.

Among Porter's smaller projects were the Schmidt cameras for Palomar Mountain. The eighteen-inch instrument mentioned earlier was completed and installed on the mountain by 1936 and put into operation by September of that year. At last some of the eager astronomers could begin to see results even if they were only reconnaissance for the main telescope to come later. Soon two more Schmidt cameras would be added to the ranks, a forty-eight-inch and an eight-inch, the latter designed by Porter and put into operation in 1940.

In April 1936, the 200-inch disc finally arrived in Pasadena from Corning, New York, and was transferred to the optical shop. Now began the gargantuan task of shaping the big disc into an astronomical mirror of the highest quality, a job that would end years later in October 1947.

Because of an accident in Corning the mirror had been made two inches thinner than planned. Although it was no thinner than theoretically required, there was a certain amount of apprehension over its figure-holding ability, which could not be ascertained until the disc was ground and polished to a sphere for optical testing. Two years passed before the surface of the mirror was ground out and polished to a sphere. The first men to examine the shape of the mirror were John Anderson, Marcus H. Brown, the self-taught man in charge of the optical team in the shop, and Russell Porter. The surface was tested with the Foucault knife-edge test, and much to everyone's relief, it showed not a trace of surface distortion. One more hurdle had been overcome. It was now assured that the glass could be turned into the working heart of the telescope.

During this same period, after most of the details of the mounting had been worked out, a one-tenth scale model was constructed, complete with mirrors. It was used for testing the various schemes, which could only be unproven ideas as long as they remained on paper. Porter and Anderson were involved in the testing of this small telescope erected on the roof of the astrophysical laboratory.

Some of the men most intimately involved with the 200-inch telescope did not live to see it completed. In February 1938 Dr. Hale died, never to see his grandest dream become a reality, but not before he was confident of its ultimate success. And in the following May Dr. Pease died.

If the Hale Telescope was dedicated as an enduring monument and tribute to one of the greatest of American astronomers, so the cutaway drawings of the great telescope by Porter live after him as his monument. Even when the telescope had been put into service and after Porter's death, the drawings were still being used instead of the blueprints to find out where things were.

Porter's technique for making the cutaway pencil drawings had been developed years earlier while he was studying under Despradelle at the Massachusetts Institute of Technology. The mechanical parts of the drawing, the lines showing edges and boundaries, were done with hard pencil. These were then nearly obliterated by shaving off particles of pencil lead with a knife or razor blade and rubbing them all over the drawing to get a gray tone. The darker shades were added next with a soft pencil, 2B to 6B. Finally the highlights were picked out with a sharpened eraser. To achieve the brightest highlights of the drawing Porter added touches of white paint. The drawings were completed with a considerable amount of intentional smudging with a well-soiled thumb to create the smudgy look of real structures.

These and other drawings Porter created from blueprints or sketches or just from verbal descriptions are considered by many to be his most important contribution. Certainly they were used at all stages of design to help clarify ideas that could best be understood in three dimensions. It is unlikely that another artist could have been found who would have been as well equipped as Porter for this task. He understood the entire project, and he understood the mechanical principles he was sketching. He had been designing telescopes for years and was quick to grasp new ideas. His mind was always waiting for his pencil to catch up. He could dash off a quick sketch to settle a misunderstanding or spend days and weeks on a detailed drawing. In all, Porter made over one thousand drawings for the 200-inch telescope and related instruments and buildings.

The more formal and complete of these drawings were published in 1947 in a booklet called *Photographic Giants of Palomar*. After examining them, a renowned American artist had this to say:

> Every once in a while a body would like to go out and tell the world something. I had this feeling when looking at the remarkable drawings by Russell Porter for the great 200″ telescope now in course of construction on top of Palomar Mountain in California.
>
> If these drawings had been made from the telescope and its machinery after it had been erected they would have been

of exceptional excellence, giving an uncanny sense of reality, with shadows accurately cast and well nigh perfect perspective: but to think that any artist had his pictorial imagination in such working order as to construct these pictures with no other material data than blue prints of plan and elevation of the various intricate forms—is simply beyond belief.

These drawings should be in a government museum of standards, in a glass case, along with the platinum pound weight, yard stick, etc. to show the world and what comes after just what a mechanical drawing should be.

Not only that, but the rendering is a work of art, exact and life like, and done with a delightful freedom of technique. They have a decided decorative quality, and in an appropriate setting could very well serve as unique mural decorations.

I doubt if there are drawings anywhere which can in any way compare with these for perfection in showing what a stupendous piece of machinery is going to look like when finished.

The pity is that comparatively few will know about them, for their creation should be world news.

Maxfield Parrish:[9]

Cornish: New Hampshire   December 15th. 1938

When the massive effort of World War II came, Cal Tech joined at once, and the 200-inch telescope had to be put aside. The men working on the project, including Porter, found themselves joining the fight, and as a result the Hale Telescope did not reach completion until 1948.

XIII. ART, SUNDIALS, AND MUSIC

Oscar Marshall once wrote that had Russell Porter so wished, he could have made fame and fortune as an artist. But except in his technical drawings for Cal Tech and for the Navy during World War II, Porter never tried to be anything but an amateur.

John Pierce once said that Porter's pencil sketches were "so real they had color."[1] He used guide lines, then started in the upper-left corner and worked to the lower right, putting in all the tonal values in one pass to avoid smudging. He drew bordering lines free hand straight and parallel without using a mechanical guide, and when he lettered free hand, the line came out centered. Pierce's lasting impression of Porter was that whenever he was not talking he was sketching.

The play of ever-changing natural color on any scene never ceased to fascinate Porter. It was in the Arctic that he first captured the mood of a scene in paintings. He painted icebergs, shorelines, glaciers, ships, and explorers' camps. He recorded men on walrus hunts, Eskimos slaughtering the vanquished, and Eskimos paddling their kayaks.

One of the most beautiful paintings of Mount McKinley ever done is by Porter. In it the valleys and lower peaks surround the soaring blue-white mass, and among the lower peaks are thin clouds that seem to be suspended in the third dimension. The painting was done for the frontispiece of Cook's book, *To the Top of the Continent,* which also included two sketches drawn from Cook's description of his theoretical final route to the summit.

For his unpublished manuscript *Arctic Fever*, Porter had made scores of sketches. In the early 1930s, during the depth of the Depression, Ingalls was unable to interest a single publisher in either text or illustrations. But the manuscript attests to Porter's wish—almost a need—to record impressions and feelings as well as people and places.

In addition to pencil, Porter worked in water colors and pastels. He did do some painting with oils, although they did not hold his interest. In the Arctic and while living in Port Clyde and Springfield, he used water colors for the most part. Then after moving to California he spent more of his time with pastels.

There was hardly a phase of his life that Porter did not record with pencil, brush, or pastel as a photographer might do with his camera. With anyone who watched him happily at work at his easel there remained a lasting impression of his abilities and techniques and most importantly of the fun he had. Dr. Knowles Ryerson was one of these persons.

Ryerson was a friend of the Porters who had grown up around the Mount Wilson astronomers and caught the fever of making telescopes. He later went on to head up the University of California at Davis. When asked if Porter ever compared the Arctic with the desert he said he "never heard him specifically try to relate his attraction to the Arctic to that of the desert." They both "offered him wide horizons and colorful landscapes." He simply "enjoyed the wide-open outdoors wherever he found it, and was thrilled by the variation that he met."[2]

Although drawing was his first love, Porter was not averse to trying other art forms. His Garden Telescope with its graceful bowl of lotus leaves exemplifies his interest not only in astronomy but in sculpture. His sundials provide other examples.

Porter's fascination with timekeeping, which had begun in his arctic days and had led to astronomy and telescope making, brought him logically to sundialing. Sundials had no practical use during most of his life, but because he was attracted to them, and because he was a "do-it-yourself" experimenter, he had to make them.

His first sundial demonstrates once again his ability to combine his scientific with his artistic hobbies. It took the form of a playful cast-copper dolphin splashing in a tiny birdbath. Facing south with his tail curled up over his head, the little dolphin balanced the wire gnomon between his nose and tail. The hour circle was supported on his back and had a built-in adjustment for the equation of time, the difference in right ascension between the real sun and the fictitious sun that moves uniformly along the celestial equator. In this way the local time could be read directly. The dolphin, with large head and small body, completely took up the space in the shell-shaped birdbath. That perhaps was unfortunate for the birds, but actually the water was only intended to provide a handy reference for leveling the sundial. In addition, the dolphin rested in a spherical mount, which allowed adjustment of the gnomon for the latitude.

Porter also enjoyed writing about the fun he had with his astronomical instrument making, and the August 1928 *Scientific Ameri-*

Porter's dolphin sundial. The hour circle is missing.

The glass-sphere sundial built by Porter while in California

can carried an article in which he described his early adventures in sundialing. "My own interest in sun dials has been aroused in the hope that in some way, perhaps by the aid of lenses, the time can be read to an accuracy of seconds instead of minutes."[3]

Borrowing an idea that had been tried as early as the 1500s he tinkered with a beam of light instead of shadow to indicate the time. But a beam of light still has a blurry edge, so in his next two dials he incorporated lenses and prisms to form a sharp image of the sun on a graduated scale. In the article he explained that the result "resembles a surveyor's transit tipped over until its horizontal plate is parallel with the plane of the earth's equator."[4] The sun's image was projected down onto the hour circle. Then by looking through an eyepiece the time could be read simultaneously while lining up the sun's image with the crosshairs. A second sundial was more elaborate. Through a series of lenses and prisms the sun's image was projected onto a ground glass along with a segment of the hour circle so the time could be read directly from the ground glass.

Perhaps the most interesting of Porter's later sundials were those he devised out of glass spheres. He bought an eleven-liter spherical flask made by Corning and mounted it equatorically with a thrust bearing and two support pads much like the mounting he had proposed for large telescopes. The hour circle was glued onto the outside of the sphere while the analemma, a metal strip carrying a plot of the equation of time, was inside. By turning the sphere the sun's image, formed by a lens placed in the neck of the sphere, fell onto the equation of time. The correct time was then read from the hour circle, estimating to the nearest minute. The glass sphere was not without its merit. A quick wipe with a cloth removed any evidence of passing birds that seem always to plague sundials. Borrowing an idea from an Australian astronomer, Porter next set to work designing and building sunclocks, sundials that show the time on a clock face. By turning a knob the user makes the sun's image fall on the analemma. Simultaneously through a gear train the arms of the clock face automatically indicate the time.

His early sunclocks were castings and, like the dolphin and the Garden Telescope, were artistically designed around leaves or flower petals to serve as garden ornaments as well as timepieces. Others had more emphasis on precision and dispensed with the flower work. He realized that the sunclock had the potential, when coupled with lenses to form sharp images of the sun, to transform the sundial into a precise timepiece. His most precise sunclocks could indicate local time to within a few seconds, always a source of amazement to visitors. According to Ingalls' report, Porter made at least nineteen sundials and sunclocks while living in California. Some of these designs were described in another article he prepared for *Scientific American* (August 1935).

Porter and Fred B. Ferson, of Biloxi, Mississippi, were friends. Ferson once made a sunclock from a Porter design, but more important things were to come of their mutual interests. The two men may have met first during one of the early conventions at Stellafane, for Ferson was just as eager a TN as Porter. One of his special passions was pouring his own metal castings, and one of Porter's enthusiasms was carving wooden figurines. These two hobbies they combined in an unusual partnership.

One summer, when Porter was free of his duties at Cal Tech, he went to Biloxi to work with Ferson making cast telescope models and figurines. Porter carved and whittled wooden patterns, and both

One of Porter's sunclocks built while he lived in California

One of the TN castings made by Ferson and Porter

*At the home of Fred B. Ferson in Biloxi, Mississippi, Porter watching
over the furnace as the two men make the TN castings
    Date unknown*

men made the castings in Ferson's backyard furnace. One of their early creations was a replica of the 200-inch dome, a brass structure with a base eight and one-half inches in diameter and a revolving dome of aluminum with attached shutter. The dome lifted off and the inside was hollow. At least one of the models had a tiny moving replica of the telescope itself. Ingalls proudly described the work to his followers in *Scientific American,* but failed to mention how many were made. Three are known still to exist.

From soft pine Porter whittled a series of four- to five-inch statuettes of the telescope maker intent on various stages of his instrument. For one he seated the hobbyist in front of his polishing bench examining his mirror pits with the concentration of a diamond cutter. Another, dated 1940, depicted the amateur testing his mirror with the knife-edge and kerosene lamp while sitting on the floor at one end of a bench that looked as though it had barely survived the long

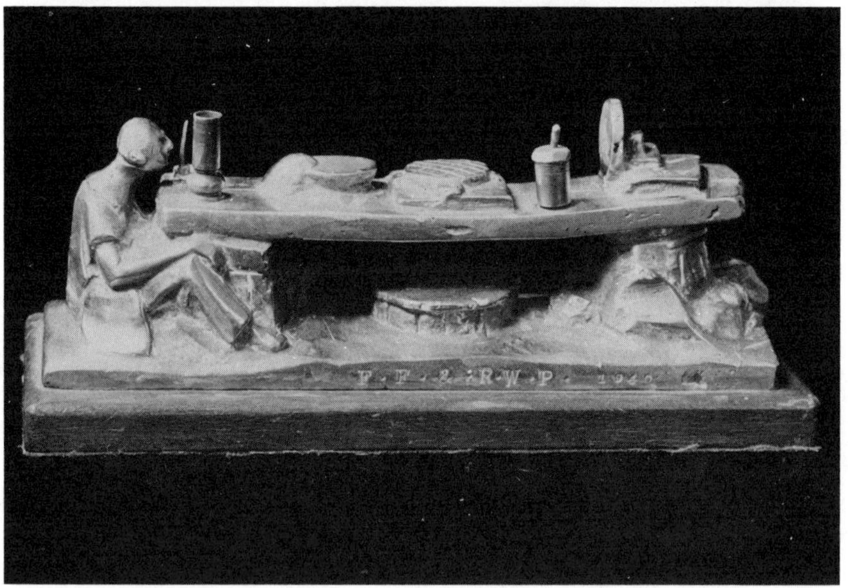

"The Mirror Maker," made by Ferson and Porter in 1940

polishing seige. In 1939 he had done a little TN absorbed in the business of directing his reflector, perhaps finished only that day, toward a celestial show. The telescope had a bent tube. Yet another had an observer lying on his back, his legs wrapped around a rubbish can to support a refracting telescope pointed skyward.

Porter's artistic sense of humor was not confined to his figurines. It often flowed from his drawing pencil as well. The subject that he enjoyed cartooning the most, as might be expected, was the one closest to his heart, telescope making, and he used it in numerous comic strips.

One showed two friends setting up a telescope to observe a comet. Soon they get into an argument over the comet's mystery, the argument boils into a fist fight, and the telescope goes crashing to the ground. Realizing that the telescope had been destroyed in the skirmish, the men decide that two friends really do not need a telescope, and they sit down on the broken tube to spend the rest of the evening pointing out the stars to each other.

Another comic strip revealed the pitfall of demanding excessive magnification from a telescope. It showed an amateur setting up his instrument at 5:00 in the afternoon. After a good deal of struggling he attempts to view the moon with 400 magnification. The image is so horribly warped and blurred that no detail can be seen. The amateur becomes so frustrated that by midnight he has taken to his telescope with an axe and completely done it in.

As he had sketched on camping trips in the Vermont mountains so Porter recorded his camping trips into the western desert, and sometimes cartooned amusing episodes. For example, he jestingly recorded the time a can of beans exploded because someone had forgotten to open it before putting it on the fire. And once he drew his camping friends shaking out their boots at daybreak to make sure no snakes had curled up inside. He was afraid of rattlesnakes in the desert since he could not hear them.

Porter's usually quiet sense of humor occasionally broke out in other ways. Although Oscar Marshall once wrote in *Popular Astronomy* that Porter was "colorblind" to poetry and that Hale had once sent him a favorite poem by Shelley to no avail, Porter was not above penning a few lines of his own. This sample was printed in *Scientific American,* or Sciam, as it is affectionately known to its readers:

1
2
3
4
5
6
R.W.P. 36

Porter cartoons, pencil and pen, 1936

> Si am de man
>   Who t'inks he can
> Make 'scopes from
>   Readin' ol' Sciam.
> She threw up her hands
>   And let out a "Damn!"
> "You've ruined my kitchen.
> A curse on Sciam."   5

If Porter had little interest in poetry, he made up for the deficiency by his interest in music. During his first arctic trip to Greenland with Cook he had watched the Eskimos dancing "round and round over the oily floor of an old warehouse." The scuffing of the sealskin boots, "the half wild music of the violins, the midnight revelry in broad daylight"[6] all made an unforgettable scene. Later when he had returned to civilization for good, he read books on harmony and counterpoint and composed several string quartets, one of them intended to recreate the rhythm and scuffing of the Eskimo boots.

During the Springfield years he experimented with composition and orchestration. When his pieces were played by a small orchestra, Porter realized how amateurish they were. But he still enjoyed his composing. It was fun, and, after all, his hobbies did not have to be widely acclaimed.

Although Porter usually avoided community affairs, he was elected director of the Springfield Musical Association in the 1920s. Its prime concern was to bring concerts to the culturally isolated villagers. On the evening of November 12, 1926, the audience heard Mario Cappelli with a piano accompaniment by Chester Cook from the Boston area. Whether Porter and Cook first met during this concert is not certain, but Cook was interested in making telescopes, and the two men formed a lasting friendship. "Chet" made numerous trips to the Porter home in Springfield to talk about music and telescope making.

Rimsky-Korsakov, Tchaikovsky, Brahms, and Mozart were Porter's idols, but Beethoven towered above all. In Pasadena Porter purchased a phonograph so he could hear his favorite works. In his bedroom, within reach of the bed, he kept the scores to Beethoven's nine symphonies and all his quartets and sonatas so he could pull them out to follow the music as the records were played. Or he could have just as much fun reading through the scores without playing the records. He could hear them either way. Porter also kept a small five-octave piano in the bedroom, on which he composed at odd moments.

At the home of one of their friends, the Porters often heard a trio, rounded up to play the works of Beethoven and other composers. If the other composers did not please him, Porter sometimes turned off his hearing aid, the one advantage, perhaps, of his deafness.

During the summer of 1939 on one of his trips east, Porter spent several days with the Ingalls in New Jersey. He forgot all about telescopes and glass grinding and withdrew to the cellar for an orgy of Beethoven listening. For twenty-three hours over a span of three days Ingalls kept his guest fortified with Beethoven records. In the seclusion Porter could control the volume to overcome his deafness without offending anyone. His favorite piece was the andante of Opus 97, the Archduke trio.

In 1944 Porter wrote Lillian Sievers, an admiring pianist and amateur astronomer, "Ingalls and I spend almost as much of our time over Beethoven as we do on optics." Ingalls favored Beethoven's sonatas and trios, while Porter was "more intrigued over his quartettes and, of course, the nine symphonies."[7] This letter was written in reply to her inquiry about a sketch of his that appeared in the advanced volume of *Amateur Telescope Making,* edited by Ingalls. It showed an amateur seated atop a ladder straining in a most uncomfortable and precarious position while looking into his homemade telescope. Flowing across the sketch were the words and notes of the amateur astronomers' theme song, a parody on Gilbert and Sullivan's "H.M.S. Pinafore." The song was not original with Porter, having been written at the Harvard Observatory in 1879. But the drawing itself exemplifies both his scientific and his artistic pursuits.

Porter's love of travel had a new outlet after the move to California. Now there were the trips east during the free summer months, first in the Model A Ford, taking different routes across the country. In 1929 they took the southern route through Arizona. In 1932 they charted their course north through Yosemite and on to Mount Rainier and Seattle, then east to Glacier and Yellowstone National parks and the Black Hills. Their peregrinations carried

Porter cartoon, pencil, 1935

Porter cartoon, pencil, 1941

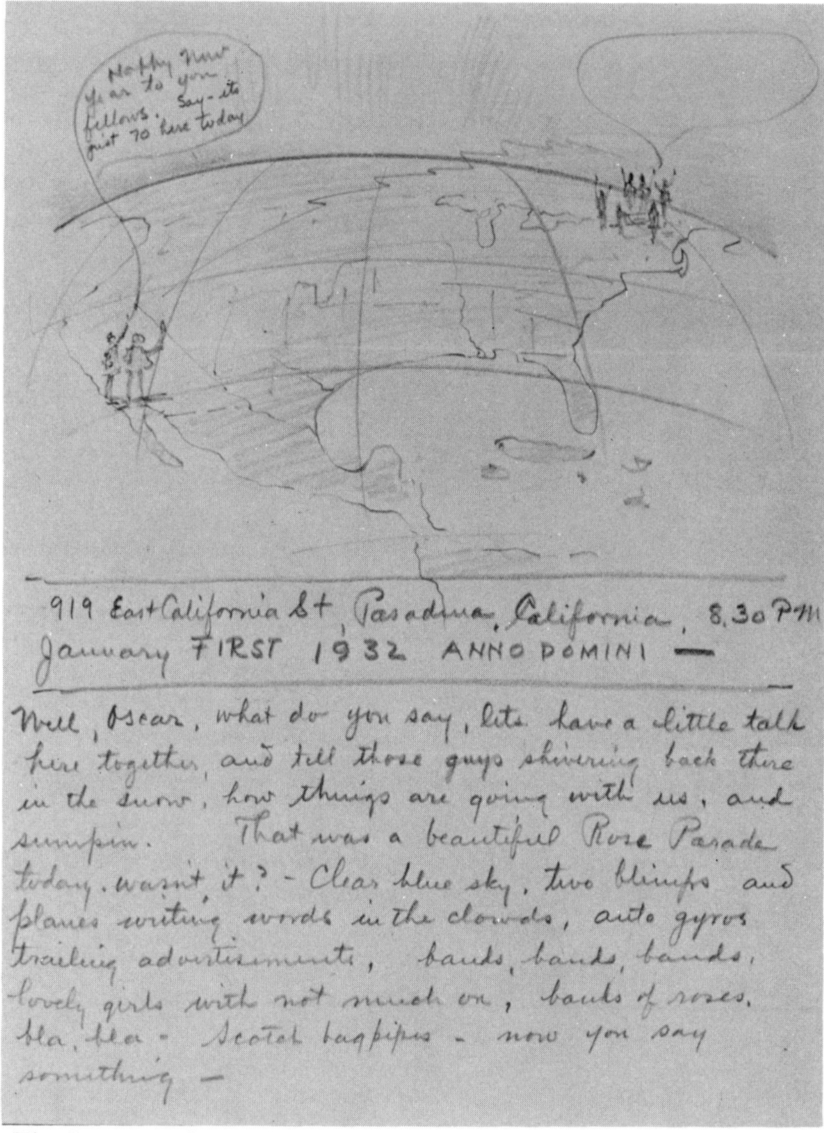

The first page of one of many letters written by Porter and Oscar Marshall to their friends in Springfield, Vermont, showing Porter's love of drawing

Cartoon of a visit to Death Valley by Porter, pencil, 1930

them to Duluth and Detroit, on to Niagara Falls, and finally to Port Clyde. They returned west by traveling south through Memphis, the Ozarks, then up to Colorado, and finally to Pasadena.

Once, after Alice and Caroline, age sixteen, had gone back to Pasadena ahead of him, Porter found a companion to drive west with him, George Perry, an engineer at Jones and Lamson. The two spent sixteen days on the road.

When Oscar Marshall was working on the Palomar project, having become a machinist there in 1929, the two families sometimes traveled together. One year they returned from Vermont on a grand tour through the southwest. In the little town of Hackberry, Arizona, Marshall took time to write the Stellafane gang about their travels. They had gone to the usual tourist sites—Rio Grande Canyon, Painted Desert, Petrified Forest, Meteor Crater, and the Grand Canyon. But their most interesting stop was at the Lowell Observatory in Flagstaff, Arizona. There they were treated to a tour of all the instruments and a viewing of hundreds of planetary photographs.

What the mountains had been in New England the desert became in California. As he had escaped Springfield on weekend camping trips, so Porter, often with Marshall, escaped Pasadena. In the desert they could camp, paint, sketch, take photographs, observe the stars, and in general enjoy life. In the fall of 1930 they visited Death Valley for the first time. The nadir of desolation was so fascinating that they returned many times over the years. Porter loved to capture with his pastels the colors he found in the ever-changing moods of the valley.

There were shorter trips, too, when work permitted, such as one taken the last week of 1931. They drove 1,300 miles through southern California and Arizona. In a letter to the gang back in Springfield written jointly by Porter and Marshall on New Year's Day 1932, Porter told of their trip. They visited old Spanish missions, walked under thirty-to-forty-foot cacti, and marveled at the ancient Indian ruin, Casa Grande, south of Phoenix. "We went over the Apache Trail and had a good look at the Roosevelt Dam." The toll of the Depression on human dignity was evident along the road. "There were a lot of desert rats—men walking along the side of the road, bed rolls on their backs." Then the two families headed south. "Of course, we had to go over the line into Mexico—now, don't

*Porter on Pike's Peak, Colorado, in 1932, during his cross-country trek
from Vermont to California*

Porter hamming the desert rat at Muroc Lake, Mojave Desert, May 1932

laugh, we didn't have a single drink. Mrs. Porter loaded up instead on old pottery."[8]

The letter went on for nine pages in a chatty fashion, alternately written by Porter and Marshall and affectionately interlaced with Porter sketches. They wrote about little things, and they wrote about the great astronomers and cosmologists on the Cal Tech campus. Porter, stopping by the office of Dr. Millikan the week before, had been shown some photographs of the recently discovered cosmic rays, taken by Millikan on a machine that Marshall had helped to build. They also wrote about the visit of Dr. Einstein to the campus. Porter finally ended the New Year's Day letter: "Yes, Oscar, its getting late and I've got to go home. There's a fellow over there with Caroline and I've got to put the skids under him. Good night. See you at the shop tomorrow."[9]

In early May 1932 Porter, Marshall, and a mutual friend, Aaron Baker, drove into the Mojave Desert to visit Muroc Dry Lake. Marshall, who was an exceptional photographer, capturing on film many incidents of the 200-inch telescope project as well as their camping trips, this time caught Porter hamming a parched victim of the desert. He titled his picture "The Poor Old Desert Rat Rescued Just As the Desert Was About to Claim Him."[10]

The three men set out again in the middle of May, this time for Death Valley. Porter wrote to a friend in Springfield, "I had been feeding them harrowing tales of those early argonauts and the heat they might expect to experience, and worrying them over the loss of our only water." When they got there it was raining. Determined not to have their spirits diluted they decided to camp anyway. That night it turned so cold they shivered the whole time, and before they got out of the valley their car was covered with mud from splashing through the pools of rain water. Porter wrote that he would not blame his companions "for putting down all those tales about Death Valley being a fiery furnace of ashes, to just plain bunk."[11]

Although Porter held a responsible position with Cal Tech on the 200-inch project, like everyone else he found it difficult to meet even the simplest of financial needs. He and Alice worked hard to get the most from a dollar. "It is surprising how little we need actual cash to get along with," he wrote in 1933. "We were caught with only a dollar or so. Most of this Mrs. Porter has won from me at solitaire." But Porter had a gas card and "always found food on the table at

Rain in Death Valley, Porter cartoon, pencil, 1932

meal times." He went on to explain Alice's uses of one dollar. "As days went by and no food shortage, I wondered how that dollar could buy so much until she told me that her pet butcher was 'carrying her along.' " For groceries Alice had a charge account and used the telephone. "If worse comes to worse we have the Atheneum (Institute Club) and an indefinite meal ticket." [12]

Not too long after this, a number of men with a mutual interest in scientific and astronomical curiosities started to meet from time to time. Some of them had been bitten by the bug of amateur telescope making, among them Porter, Marshall, Byron Graves, and James Barkelew, a patent attorney and friend of Porter. Some, like John Strong, Robert Leighton, and Rudolph Langer, were physicists at Cal Tech. There was a biologist, a chemist, some Mount Wilson astronomers including Dr. Anderson and Dr. Enrique Gaviola, inventor of the caustic knife-edge test for mirrors, who was working with Dr. Strong.

The original concept of these meetings was to deliberate over theories and ideas, mostly in the field of science, that did not have one chance in a hundred of being proved or discredited. One of the early problems they discussed was how could the principle of the Foucault pendulum be employed in other ways to demonstrate that the earth is a rotating body. Dr. Anderson proposed the name One Hundred-To-One-Shot Club and thereafter that is what they were called.

Once a year, usually in June, they took an overnight trip to Palomar Mountain, camping in a clump of oak trees a few hundred yards south of the 200-inch dome. On the first trip they saw the holes in the ground for the dome foundations. During the last trip they saw Jupiter and the Hercules Cluster at the Coudé focus of the new giant.

When the weather was good the monthly meetings were supplemented by weekend trips into the desert. The men seemed to take great pleasure in rolling up their sleeves and having a good time with what was to many of them their vocation. They generally took a telescope along on these camping trips to view the stars in a dark desert night. Sometimes they tried to find Venus during the day. Graves, a large man with a gentle smile, wearing a wide-brimmed hat, suspenders, and plaid shirt, took over as cook, and often Strong became his assistant. Of course Porter used the opportunities to sketch or paint the desert.

The One-Hundred-To-One-Shot Club on a weekend trip to a cabin in the desert, circa 1934, viewing Venus in the daytime with a refractor designed by Porter
Left to right: Byron Graves, Russell Porter, Oscar Marshall, James Barkelew, Stan Hart, Dr. John Strong

On one trip into the Mojave Desert the men discovered a vacant stone cabin. A sign indicated that anyone was welcome to use it but please leave it as found. The cabin soon became a favorite camping site, and the Hundred-To-One-Shot Club returned many times. Out of gratitude for the use of the cabin they reshingled the leaky roof on one trip. Another time they built a dutch oven. They never met the owner.

Years later Dr. Strong recalled the trip to the stone cabin during which they decided to find out where it was, geographically speaking. Of course they had failed to bring a sextant or theodolite and so decided to face the challenge squarely and make do with what was on hand, which was not much. Someone remembered enough about the equation of time to estimate how much the apparent sun was leading or lagging the mean sun for that date. Relying on a watch for local mean time, they determined when the sun crossed the meridian by noting when a shadow had its shortest length. Knowing what hour meridian corresponded to their local mean time, they determined the longitude. Someone found a piece of string from the grocery supplies, and with this and a plumb bob the necessary measurements were made of the sun's altitude as it crossed the meridian. One of the group, a mathematician, wrote down the expansion series for the cosine of an angle, then used the cosine of the sun's altitude to come up with its altitude. After estimating the sun's declination for that date, the second coordinate for the stone cabin was found, its latitude. Later the group found that they had put the cabin on the map within four or five miles of its actual location.

So many of Porter's activities show his warm enjoyment of companionship that it is not surprising that he had the habit of bringing home unexpected guests. Alice might or might not know them, but she was accustomed to making them welcome at all hours of the day and night. She enjoyed having guests in her home, and she shared the pleasure that Russell took in having musical or astronomical gatherings. Throughout their life together, Alice unselfishly shared her husband with countless friends and strangers, a thoroughly devoted wife. Cold dinners when Porter worked late, predawn breakfasts to get weekend trips underway, late snacks and coffee during all night star gazing, she provided for these occasions with an undiminished sense of duty.

Outside of the sphere of her husband's influence Alice had her

Porter at an old Hollywood site in the desert, 1930

own interests of which the favorite was horticulture. She spent countless hours in her flower gardens and eagerly attended flower shows. Both in Springfield and in Pasadena her gardens were the envy of friends and neighbors, and her living pattern gave her the opportunity to study and enjoy the flowers indigenous to two very different regions.

Plagued by ill health during 1934, she was unable to join Russell in his weekend excursions, but at the beginning of 1935 was well enough so that they could visit the construction site of Boulder Dam. "We were there when work was started 2–3 years ago," Porter wrote, "and were anxious to see the finished dam before they closed the gates and started the hundred mile lake." While there they enjoyed taking a guided tour and standing at the foot of the dam on the upstream side and trying to imagine what it would be like when the water reached the fill mark. "My biggest kick was seeing them roll those 3″ steel plates into 30 ft. tubes." After welding them "they hook 'em up to a couple of tractors, one ahead one behind, and go hell bent down a stiff grade to where the cable trolleys carry em out and down to the dam or power house."[13]

If his fancy could be tickled by such a feat of engineering so could it be by quite different entertainment. Once while visiting the desert near Yuma, Arizona, he and some friends stopped to inspect the old site used in the filming of "Beau Geste" and "Desert Song." His boyish antics bubbled to the surface when he posed for photographs in various positions. In one he is holding a dead palm branch as if it were growing out of the top of his head.

Later he went to Hollywood and took the opportunity to visit the studios of Metro-Goldwyn-Mayer. He got a youthful thrill out of his special chance to meet Cecil B. DeMille. Then he inspected sound recording equipment and ate in the studio restaurant where he could watch the actors and actresses dressed in an endless variety of costumes.

His roguish, boyish grin and the mischievous twinkle in his blue eyes showed that he never really grew up, and he was always ready to smile. Porter was a man who needed to be constantly busy at something, but his great talent for combining his job and his hobbies, his work and his leisure, gave his varied and enthusiastic life a remarkable unity.

Russell, Alice, and Caroline Porter on Mount Wilson, October 1929

*Caroline Porter and a younger cousin in California
Date unknown*

Porter with his pastels in California
Date unknown

XIV. AMATEURS AND HONORS

When Porter left for Pasadena in 1928, he left Vice-president Pierce in charge of the Springfield Telescope Makers, expecting the next summer to pick up where he had left off. Instead, the move was to become permanent, and his association with the group was to change over the years. He remained active, however, attended most of the Stellafane conventions, and was in constant communication with club members. His work for Cal Tech had no influence on his deep feelings for these men with whom he had shared so many happy and exciting hours. And, of course, his progress reports on the Palomar project were eagerly welcomed.

In 1929, 110 people signed the register for the August 10–11 convention at Stellafane. Many of them were Porter's old friends. During the day he examined telescopes brought up the hill and tested several mirrors made by the amateurs. After supper everyone listened intently to his first report on the 200-inch-telescope plans and his contribution to the project. He said it would take ten year's effort to build the giant instrument; it ended up taking twenty. His main work so far, he added, was in designing the machine and optical shop buildings.

On September 6 Porter attended the regular monthly meeting of the telescope makers group, bringing plans for a large telescope to be erected in front of the clubhouse on a rock outcropping. Some of the parts of the equatorial mounting were familiar to certain members because they had been machined at the Jones and Lamson

The reflecting turret telescope being erected at Stellafane, 1930, with Porter on the right in coveralls

*The right ascension ring being hoisted into position on the turret telescope
at Stellafane, 1930, with Porter in coveralls and a white hat*

Machine Company between 1924 and 1926. Porter had earlier written an article in the May 1921 issue of *Popular Astronomy* in which he proposed a turret telescope modeled after Hartness's telescope. But this one would use reflecting optics instead of lenses.

Now that his observatory at Port Clyde was no longer used, the sixteen-inch mirrors that Porter had made were available for another telescope. They provided the very stimulus needed to bond the entire club to a new project, the first reflecting turret telescope in the world. Porter planted the idea, provided some plans, and then had to leave for California. But construction was started that fall.

Years later, John C. Pierce recalled that the September meeting had been for him an early experience with Porter's character. Porter had been describing the pendulum experiment of measuring the time for each swing and plotting this as a function of the pendulum's length. Pierce, son of the club's president and, at fifteen, the youngest person present told how he had done the experiment and obtained a nearly straight line plot. Porter let this go for just a short time and then quietly assured him that it was really half a parabola. Even with all the other older men present, John was not made to feel embarrassed. Porter did not talk down to him.

The next summer, all hands, including Porter, gathered on the hilltop to raise the stellar fortress. The wooden forms for the concrete had already been constructed just north of the clubhouse. Nearby on the ground lay the steel equatorial ring and seventeen-foot boom that would later jut from the turret to hold the parabolic mirror. A pile of sand, bags of cement, and barrels of water stood ready for the task. The other men arrived in their work clothes and coveralls. Porter appeared in his knickers with stogie blazing. Before the day ended countless shovels full of sand and cement had been tossed into the small cement mixer. No one bothered to keep track of how many loads were emptied from the mixer and dumped into the forms. Everyone was too busy hauling water, shoveling sand, pouring concrete, and sweating under the summer sun. But they labored for the love of it; they were erecting a monument of concrete, steel, and glass.

After the walls had set they returned and cautiously skidded the seven-foot-diameter steel equatorial mount up a heavy wooden-beam ramp and secured it in its final position. Oscar Marshall carefully recorded each step with his "five-by-seven" camera. The men look happy and satisfied, but the Vermonters were not about to waste any smiles or cheers while work still remained to be done.

The partially completed turret telescope was the main attraction during the fifth annual convention at Stellafane that August. The work continued after the Porters' return to Pasadena in October, but was not completed until the following summer.

That year the host club had a special drawing card for the visiting TNs: their new telescope of sixteen-inch aperture. It had been hurried into working order just in time for the convention even though the motor drive was not rigged up until later. It must have been a proud moment for Porter to see the fine product of all the hard labor of his followers and friends. At last the mirrors he had made so many years ago at Port Clyde were being used again. Over a hundred registered guests were present to admire and envy the new turret telescope. Very few of the amateurs among them had the use of so large a telescope.

Not too long after the Porters returned to Pasadena, Caroline suddenly eloped from the Porter home at 615 South Mentor Avenue. What a shock it must have been to her parents to realize that their daughter chose elopement to the usual wedding. There was little communication between parents and daughter after that. In all of his subsequent letters to his friends at Stellafane Porter never mentioned Caroline. Alice and Russell rarely talked of her to friends or visitors although they kept a photograph of her in the living room. Little else is known about the situation until after World War II, when we hear of a visit to Caroline in Maine.

Porter almost never mentioned politics either. He was always more interested in discussing scientific or technical problems, but occasionally a letter would contain some kidding political overtures. From a person raised in the Republican state of Vermont the following remarks, written after the 1932 national elections, are not surprising: "I suppose you are all watching for the return of the Democratic party to Power." He went on to say how they had "one of that color in our organization [Stellafane] and yon might ask him to let us know just what day prosperity will arrive." [1]

The summer of 1934 was a disappointing one for Porter. "This is the first time I have failed to be with you since the movement was founded ten years ago," Porter wrote to the amateurs gathered at Stellafane for the annual convention, "and nothing short of abso-

The turret telescope at Stellafane just after completion in 1930

lute necessity prevents my attendance this year." The pace of the work on the 200-inch telescope was increasing, "and it really looks as though I might get a look through the thing before I check out." Yet the most important reason for not coming east was "Mrs. Porter has been ill for a good deal of the year but is slowly improving." [2]

The next year it was Porter's health that kept him in California. In the spring he had been stricken with a serious heart attack. He was sixty-three years old. Ingalls, in the November *Scientific American*, reported that Porter had had to spend five months at home recuperating. But even while bedridden and inactive, he could not stop his mind, which turned over and over with new ideas, and he finished a chapter on telescope mountings for a new edition of *Amateur Telescope Making*. He fully recovered, however, and was able to resume his old job and activities, going back to Stellafane conventions in later years.

Although these annual treks were important to him, they were not the most important manifestation of involvement with struggling neophytes. His articles in *Scientific American* and other magazines reached, encouraged, and helped far more people. His contributions to *Amateur Telescope Making*, which was later expanded to two and then three volumes, seem almost endless. He showed great understanding and sympathy for the person who wanted to make his own telescope but who had little spare cash and even less facility with machines. Through his writing he showed the amateurs how to make mountings and tubes from simple materials that did not require expensive machining. He encouraged them to seach the junk yards and scrounge parts from old cars, washing machines, and any other derelict that might serve the purpose. He told them to use their imaginations to devise ways to make parts inexpensively. But cost should not be the only guideline. He insisted that massiveness and sturdiness must not be sacrificed for cheapness. To prove the goal could be reached cheaply, he designed a fork mounting dubbed "Porter's Folly" to be made of concrete—cheap and yet so massive that there were no vibrations. It exists even today ready to thwart attempts at superior concepts.

He designed and built mirror polishing machines and small lens grinding machines, which he used successfully and then passed on his experience to the amateurs through the pages of *ATM*. When Dr. Hale, who was also interested in encouraging amateur astronomers, agreed to make contributions to *ATM* on solar observing and instrumentation, Porter helped him design a spectrohelioscope that could be made inexpensively. He also lavishly illustrated the resulting chapter with his easily understood drawings. Countless sketches were salted throughout other chapters to help authors make their points.

For all the ideas, inventions, instructions, and drawings that Porter gave to the telescope fraternity, he always remained modest. Sometimes he was too unassuming to take credit for an original idea. An example is the Springfield mounting, which he described in the first volume of *Amateur Telescope Making* but without mentioning that it was his invention. In *ATM II* he presented an improved design for the Springfield mount. Even then he did not mention that it was his invention, and Ingalls had to correct the omission in a footnote. The new design had been thoroughly tested by Porter's friend, Fred B. Ferson, before it was included in *ATM II*. Porter told how to make the patterns for the required castings, and Ferson wrote about his experiences in actually making the castings and complete telescope.

In all his writings, Porter could not conceal the pleasure he had gained in watching the telescope craze gather strength among his own band of friends at Stellafane and gain momentum as it spread out from Breezy Hill. Of his remunerations Porter once wrote to Dr. Hale, "I always turn over the returns from my sketches and articles to Mrs. Porter as her pin money—so there's no telling for what feminine adornment the money has been devoted." [3]

It may seem that Porter spent most of his efforts communicating with other telescope makers by articles and books. But he also took great pleasure in meeting people, seeing people light up with excitement in their voices and happiness in their eyes while proudly discussing or using their homemade telescopes. He was particularly defenseless against a young boy or girl determined to put together a telescope that perhaps was beyond his or her understanding. One day Porter and Marshall were driving along a street when they spotted a young lad bent intently over what appeared to be a reflecting telescope. They stopped to investigate and discovered that the boy had assembled a rudimentary telescope from a shaving mirror he had bought for twenty-five cents and a piece of looking glass mounted in a tube. The whole thing was attached to a simple and rather shaky fork mount. The telescope did not perform very well, but to Porter what was most important was the excitement and determination of that thirteen-year-old to make a real telescope. He later invited the boy over to Cal Tech and showed him how they were making a 200-inch telescope.

On another occasion, Porter hunted long and hard for a Japanese fruit grower in Imperial Valley who had built his own telescope and had used it to discover a comet. He finally found him. His thatched roof home was made of sticks and burlap bags. The middle-aged Japanese was pleasant and well spoken in English. He showed Porter the mirrors he had made and told him of the struggles he endured with his pitch lap in the summer heat. Porter inspected the two tele-

Porter examining a rudimentary telescope made by a thirteen-year-old boy in California
Date unknown

scopes and again invited a new friend to Pasadena for a visit at Cal Tech.

Once Porter joined a group of friends for an all-night observing session on Palomar Mountain. Chester Silvernail, who was with the group and who later helped polish the 200-inch mirror, reminisced about that trip on which he first met Porter. It was a moonless night, and most of the men concentrated on observing deep sky objects. The old hands like Porter were kept busy showing the others where to look. Near midnight Porter was the first one to see the Gegenschein, that faint patch of light in the sky directly opposite the sun. He also had showed the others where to look for the zodiacal light. Later, in the early morning hours he retired for some rest, rising again at six. By this time radiational cooling had lowered the temperature enough so there was a layer of ice on the water bucket. Undaunted by this inconvenience, the old Vermonter broke away the ice and washed his face and hands, to the amazement of the Southern Californians.

By this time magazine publishers were writing about Porter as the father of amateur telescope making in the United States. While he acknowledged the fact that he had had something to do with it, he was quick to suggest it was the Depression that gave people time to pursue this fascinating yet inexpensive hobby.

Porter's influence sometimes extended beyond the single amateur working alone on his telescope. An example is the Griffith Planetarium designed by the Department of Recreation and Parks of the City of Los Angeles. Porter was asked to advise on the design of the scientific displays for the Hall of Science. To this task he readily set his mind and pencil with enthusiasm and devotion. He believed in the project because it would serve to educate the public. It was more play than work as he proposed a twelve-inch Zeiss telescope, a twelve-inch solar telescope with spectroscope and spectrohelioscope, moon model, and numerous physics demonstrations. He also had an agreement to consult on architectural matters with the architects of the planetarium.

Other recognition had come his way, too. Early in 1933, he had been singled out as a member of a special committee dealing with the results of the disastrous Southern California earthquake. It began on Friday, March 10, at 5:54 P.M. Russell and Alice Porter had just sat down to dinner at home when they felt a small earth tremor. It was as though a heavy truck was driving by. They immediately went outdoors to the patio in case another shake came. In ten minutes it happened again with the same intensity as the first time. The movement was sideways and not up and down. After waiting a time, since no other tremors appeared, they decided to continue their plans to go to a movie in South Pasadena. But at the theater a third quake came, sending the small audience stampeding for the door and "I, for the third time in my life, tasted copper," Porter wrote. In fifteen minutes, another tremor, and "we decided it was no place to be in and bolted." [4]

News of the earthquake came trickling in for most of the night over the Porter radio. It finally became evident that the quake had been a major disaster: on the order of 120 persons killed, 2,500 seriously injured and 50 million dollars damage, blocks of buildings completely destroyed, broken gas mains, fires, and looting. The center of the quake had been south of the Porters at Long Beach and Huntington Beach. Pasadena received little damage.

Up on Mount Wilson the astronomers had a case of strange disappearance on their hands. Earlier the 100-inch mirror had been removed from the telescope and lowered through a hole in the floor to the silvering room. Soon after it had received a fresh coating of silver the first tremor struck. The astronomers hurried over to the telescope to check for possible damage. The silver coating was completely gone: some of the mercury from one of the bearings had spilled out during the quake and had run down onto the mirror, amalgamating every trace of the silver.

While listening to the radio for quake news, the Porters heard an appeal for food and clothing. "We rummaged in our closets and kitchen cupboards and filled the car and drove down to the San Gabriel City hall." [5]

Before the tragic weekend ended, Porter was asked to join a committee of engineers from Cal Tech to go to the disaster area to learn something about the causes of building collapse and methods of engineering earthquake-resistant structures. Porter's special talent had been recognized: The head of the engineering department, Professor Martel, "tells me my facility with my pencil in depicting structural weaknesses will be of more use to them than photographs." [6]

After returning from the Stellafane convention of 1936, he was

cited for distinguished service by the Pasadena Post, No. 13, of the
American Legion. The citation was given to those people whose lives
had been lived for the enrichment of the lives of others. In Porter's
case it was granted partly for his contributions in the field of ad-
vancing man's understanding of the universe. Secondly it was for his
unselfish and unending gift to the growing hobby of amateur
astronomy.

Porter was reported in a newspaper account to have stated that
he appreciated this recognition more than a doctorate from a university. His biggest joy was seeing how amateur telescope making
and astronomy had brought pleasure to thousands of people and
how it had created in them an interest in the universe and man's
place in it.

In 1938 when Dr. Goethe Link was visiting some of the western
observatories on his honeymoon (he was in his late fifties at the
time), he paid a call on Porter at Cal Tech. Because he was planning
to have a large observatory built in Indiana, he asked for Porter's
suggestions on how it should be constructed. Porter obliged him
with his usual effervescent enthusiasm for such topics and produced
a sketch of what he thought the observatory ought to look like.
When Link had the observatory built in Bloomington, Indiana,
Porter's sketch was followed except in the dome for the thirty-six-
inch reflector. This was made of wood instead of steel. The mirror
blank for this telescope was one of the Pyrex pourings from Corning
made during the learning process on the way to the 200-inch disc.
The thirty-six-inch reflector and a five-inch Zeiss refractor were
products of a team effort by members of the Indiana Astronomical
Society, amateurs all.

In October of 1939 Porter was invited to join the Astronomical
Society of Edinburg, Scotand. That same year a California planetary
observer suggested that a lunar crater should be named in his honor.
Ten years later, with backing from *Scientific American*, the idea was
presented to H. P. Wilkins, F.R.A.S., Director of the Lunar Section
of the British Astronomical Association. Wilkins enthusiastically accepted the proposal. He chose Clavius B, a crater about twenty-five
miles in diameter on the north wall of the prominent walled plain,
Clavius, near the south pole. This was an ideal choice, for not far
away lie the craters named for the antarctic explorers Amundsen and
Scott. Between Clavius B and the lunar south pole lies crater Newton.

He placed Porter's name on his 300-inch master map from which all future maps published by the Association would be taken. Oddly, the International Astronomical Union, the body of professional astronomers empowered to officially name lunar blemishes, did not accept the renaming of Clavius B for Porter until 1970, partly because of the nature of Wilkins' mapping.

Also in 1939, Tulane University, New Orleans, became the recipient of an astronomical gift. The granddaughter of Professor William H. Pickering, Mrs. Margaret Pickering Zemurray, gave to the University the fifteen-inch reflecting telescope used for many years in Mandeville, Jamaica, by her grandfather. This called for the construction of an observatory planned as a memorial to the late southern industrialist, Thomas F. Cunningham. It was through his friend Fred Ferson that Porter learned of this new observatory. He had a deep appreciation for the work done by Pickering in Jamaica and was interested in seeing his telescope put to use in the United States. Ferson, who had already offered to make some missing parts for the telescope, saw to it that the architect's plans were sent to Porter. These he examined one weekend in the fall of 1940 with two engineers on the 200-inch project, then promptly returned them to Tulane University with a few suggested changes attached. All his ideas were incorporated in the revised plans.

In November of that year, Alice and Russell went to Pittsburgh by train to participate in the John A. Brashear 100th-anniversary celebration. It was from Brashear's company that Hartness had ordered in 1913 the sixteen-inch mirror blanks that helped launch Porter's telescope making career. Then the Porters rode south in time to celebrate Thanksgiving with the Fersons in Biloxi. On Saturday, November 30, they were taken to see the site of the new observatory on the Tulane University campus before heading back to California.

The growth of the telescope making hobby brought a bonus to the country as a whole. When the nation finally was forced to rally to the tragedy of World War II, dozens of the more advanced amateurs were ready to test their skills at optical fabrication of prisms. They proved they could equal and even surpass the acceptance records of the professional optical companies. Porter can be credited with helping to organize this movement. Men who had started out as amateurs using Porter's instructions became in a time of national crisis true professionals.

One of Porter's drawings made for the war effort in 1943

XV. DRAWINGS AND PRISMS FOR UNCLE SAM

When World War II came and the scientists and engineers at the California Institute of Technology, like their colleagues across the land, turned their energies from peaceful research to secret war projects, Porter soon learned that his artistic talent was immeasurably valuable to them when they conferred on weapon design with "the Brass" in Washington. His three-dimensional cutaway drawings had been used successfully for the 200-inch telescope, and now the skill behind them could be put to use in support of his country's war rally. Although Porter had been a quiet and gentle man all his life, he was not averse to working on war projects and never had any doubts that his side could beat the enemy.

At first he was engaged in projects designed to meet the threat of a Japanese invasion of the west coast. Later, when it became evident that the penetration would reach no farther than Pearl Harbor, the Cal Tech efforts were switched to offensive weapons development. Considerable effort was devoted to developing rockets for the Navy. There were rockets of all kinds: rockets launched from Jeeps, from fighter planes, from torpedo boats and landing craft. Using the technique he had developed so well, Porter now studied blueprints of a mechanism and then set to work making three-dimensional drawings, cutting away just the right amount to expose the operation of the innermost parts. Thus, before a single prototype was made, the proposed weapon could be demonstrated convincingly in Washington.

Porter at his drawing board in Pasadena in 1945

Even after mechanisms had been made, his drawings were often superior to photographs, because they did not require destroying some parts to show others in their natural arrangement. During the course of the war Porter made so many of these drawings, several hundred, that he was known in Washington as "The Cutaway Drawing Man."

During the winter of 1943 the Naval Ordnance established a test station at Inyokern in the desert country north of Pasadena and east of the Sierra Nevada. Here Porter spent days at a time just drawing. He made drawings of a test site from a certain perspective and drawings of a proposed invasion landing site somewhere in the Pacific showing landing craft, enemy emplacements, and surrounding territory. In addition, there was what seemed like an endless line of rocket fuses, launchers, and other devices. Naturally much of this work had to be produced quickly, and Porter was under much pressure to have things done "by tomorrow." The man who had suffered a heart attack in 1935 and who had turned seventy within a week of the bombing of Pearl Harbor protested that he needed his rest. But his pleas amounted to as much as a desert mirage. At one point the pace so tired him that he was forced to stay in bed for several days, but even there he managed to draw. These drawings were included in a report written to show the results of establishing the Inyokern Test Station for the Naval Ordnance in coordination with Cal Tech.

Years later in *Scientific American* Albert Ingalls told how on another occasion Porter had been pushed so hard by the demands of the war that he was forced into bed for several weeks to regain his strength. But if he was failing in endurance he at least kept his sense of humor. Ingalls quoted one of Porter's letters: he was blessed with three good-looking nurses, but only the head nurse would let him hold her hand. In April 1944 Oscar Marshall wrote to the fellows in Springfield that Russell was "visibly aging, but cheerfully as you know would be the case."[1]

Porter even got involved with the Japanese balloon bombs that were causing some west coast scares. He was asked to make dissection drawings of their mechanism. In May 1945 the War Department made an official announcement that the Japanese had been sending over balloons laden with incendiary bombs, apparently intended to start forest fires along the west coast. Thirty-five feet in diameter and made of silk paper, they had been designed to skip over the Pacific from Japan by riding the air streams between 25,000 and 35,000 feet.

Their journey was controlled by barometrically released sandbags. According to the design, the balloons would reach the United States when the last sandbag fell and then the bombs would begin to release as the sandbags had. When the last bomb was released the balloon would self-destruct.

Little damage was claimed, although the newspapers did report that six people had been killed when they investigated a balloon that had landed with a dud bomb attached. It was finally decided that the public must be warned of the possible danger even if it meant revealing to the enemy valuable information about the balloon destinations.

Porter's second contribution to the war effort was more in line with his avocation. When the United States entered the war and the industrial complex hastily geared up to turn out war machines, it was soon realized that the demand for certain optical instruments could not be met fast enough. The antiaircraft and field guns that were rolling off the assembly lines required special optical sights with a right-angle bend in them. This called for a special prism of the highest precision. Thousands were needed. Some officials even ventured to say the situation was desperate.

No simple prism with a single reflecting surface would do because it would show the gunner a backward, or "left handed," image alien to his natural aiming and tracking ability. In place of the single reflecting surface the special prism had to have a double surface or "roof" so the light was reflected twice, producing a "right handed" image. So accurately must this roof be made in order not to produce double images that its right angle must not deviate from ninety degrees by more than two seconds of arc, or 1/1800th of a degree. Herein lay the problem facing the War Department. Only an optician with the greatest skill, patience, and experience could hope to produce acceptable "Amici prisms," as these little faceted gems of optical glass were called. Even the optician with long years of valuable service to his parent optical company might not have had the chance to develop the necessary skill and finesse.

So it was that the Frankford Arsenal discovered that the opticians who had been satisfying the peacetime demands for these prisms could not possibly meet the schedules of delivery thrust upon them by the war. Efforts were made to hire and train new opticians, but even so, it would take a while for supply to catch up with demand.

Some of the skilled opticians went to the Arsenal for a time to learn what they could about these Amicis.

Even before the war broke out, Porter's friend Fred Ferson had guessed that if the war did come there would be a great demand for these roof prisms. There were no instructions on how to go about producing them to the tolerances required, so he set out to teach himself the art. Relying on the skills he had acquired as an amateur telescope maker, and equipped with patience and determination, he finally succeeded in reaching his goal.

By 1940 he had become so dedicated to the problem of prism making that he had quit his regular job as an insurance agent in order to devote his entire energy to the challenge. With severely limited capital he had to design and build his own machinery as it was needed. During the period of his self-apprenticeship he gained invaluable encouragement and guidance from Porter. At the same time, Porter, through communication with Ferson, experimented with the problems of making roof prisms and learned to appreciate what was required of a craftman in order to shape the many faceted prism, grind and polish its optical surfaces to the specified dimensions, and finally bring the roof angle to within two arc seconds of perfection.

During the summer of 1941 the Porters were again in New England in time for the annual convention at Stellafane on August 2. This year a special war message was presented to the amateurs who flocked to the hilltop. Porter was not one to make rallying speeches; he was more inclined to inspire by example and let others sound the battle cry. But he did give a talk on how he believed advanced amateurs working alone could make the prism. He then went on to give his annual report of progress on the Palomar Mountain observatory. The observatory and all the other buildings were now completed. The 200-inch disc was polished and ready for parabolizing.

Dr. Harlow Shapley of the Harvard Observatory also spoke on roof prisms. He stressed the great need for trained opticians and the fact that he felt this need could be filled from the ranks of amateur telescope makers.

Albert Ingalls drove up to Vermont to be with his fellow TNs and tell them how urgently their skills were needed, how Porter and Ferson had already experimented with making the Amici prisms and now planned to prepare a report of their work. Ferson even came up from Mississippi to talk about his experience. He went into some detail about the way in which the amateur could set himself up to make these optical gems. The movement had been started.

Besides talking with the amateurs, Porter, Ingalls, and Ferson had to convince the military command that the more advanced men, who could make telescope optics the equal of many professional products, could make the roof prisms and were eager to give it a try. With this goal in mind, The Three Musketeers, as Ingalls dubbed them, made a trip to the Frankfort Arsenal in Philadelphia.

There was much doubt in their minds about the outcome of the meeting. Could they present a convincing story? Would they be allowed a chance to see if the amateurs could make the prisms? They could bring little evidence to support their convictions. The fact that Ferson had already had some of his prisms accepted by the Arsenal was all they had to show. Of their efforts, Ingalls wrote afterward: "Would the quest prove quixotic? Would the Musketeers be laughed at—amateurs, of all people, aspiring to make roof prisms, of all things!"[2] But the officials at the Arsenal did listen. After all they had to meet the production demand somehow.

The three expounded on Ferson's prism successes and explained that before this all he had done was make three telescopes, three optical flats, and several small lenses. Hundreds of other men had similar experience, and many of them should be able to take the next step. Of course to tool up and take this next step would require the necessary glass, which was scarce. The Arsenal was not about to throw any away.

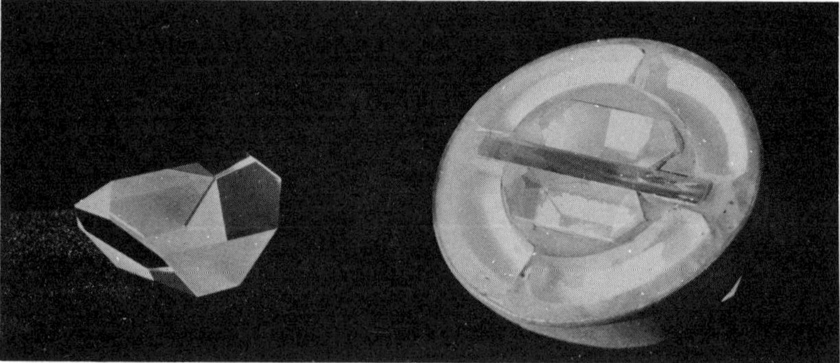

One Amici prism, left, and two Amici prisms, right, blocked for polishing the roof angle

But the amateurs did not ask for large quantities of the precious glass or for big contracts. They simply requested a small quantity of the optical glass and an agreement that the prisms would be tested and rated for quality by the Arsenal. If the results were acceptable, small trial contracts would then be awarded to the most promising producers. The strategy worked, their modest requests were granted, and The Three Musketeers went home, happily confronted with the new challenge of rallying the amateurs to the cause. Ingalls agreed to act as a clearing house, "a sort of old mother hen for the venture." [3] He lined up about eighty enthusiasts who seemed likely to succeed and mailed each man two glass blanks, treating them as if they were the last round of fresh water on a crowded life raft. Later he received the finished prisms and sent them on to the Frankford Arsenal for evaluation.

From the very beginning "The Gang," as the scattered band came to be called, was made up of middle-aged men who already had full-time jobs and could make prisms only in their spare time as they had made their telescopes. Because they had been selected from the ranks of amateur telescope makers, they naturally represented the same cross-section of the American scene: biologist, steel worker, candy manufacturer, physicist, geologist, cabinet maker, gravestone manufacturer, dentist, and so on.

The most pressing problem at the beginning was to get instructions to The Gang. In California, Porter wrote a booklet describing the testing procedures to determine if the roof angle was accurate to within 1/1800th of one degree. Ferson wrote a set of instructions for making prisms. He gleaned his facts from all he had learned during his short experience. Sixteen sketches by Porter were used for illustration. One of the amateurs had worked previously at the Frankford Arsenal as a professional and was thus able to contribute invaluably to this collection of information.

Because the members of The Gang were spread throughout the country, they could not meet to discuss their problems. To overcome this difficulty they relied on the mail. Again Ingalls handled the flood of letters, answering questions, passing on information, encouraging, and keeping track of everyone's progress. He then passed on the letters to the other two Musketeers so they could keep track of what was going on and add their own contributions. Many were the vocations of The Gang that were drawn on to help solve the problems as

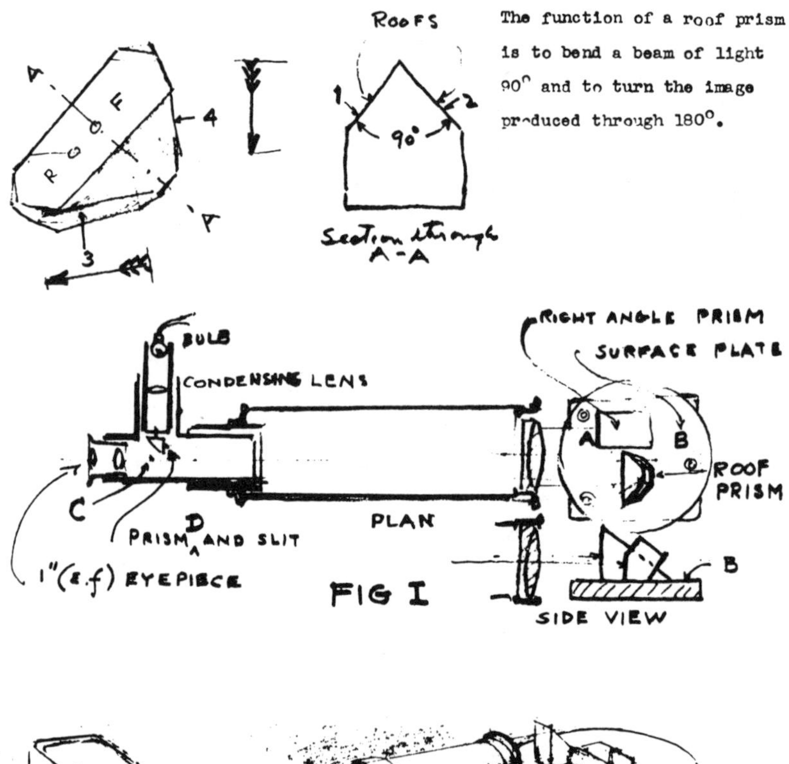

Part of the manual about testing Amici (roof) prisms that Porter wrote for the amateurs who made them for the war effort

they came up. Everyone helped everyone through the continual flow of letters and mimeographed news bulletins.

It cost each amateur from $100 to $200 to tool up for the prism making, a fact that pleased them because this was a fraction of the cost to a professional. They already had much of the necessary machinery from their telescope making activities and what they did not have, like polishing spindles and autocollimators, they could make or scrounge from a scrap dealer. Their equipment did not have to look impressive; it only had to be functional. And even before they started the job, they had a flair for operating in spartan shops and making what meager tools were needed. To a TN, part of the fun of the telescope hobby is seeing how much he can make with how few tools. He never tires of bragging to other amateurs about how little he spent on a project.

Some weeks after the men tooled up, the first prism appeared on Ingalls' desk, wrapped in a mountain of insulation. Ingalls wrote that "the first one seemed like a maharaja's gem."[4] It was sent on to the Arsenal for testing. To the disappointment of everyone involved, it was out of tolerance. The roof angle was not accurate enough. This was a major setback for the amateurs. There was nothing to do except wait and hope the next samples would be better. Soon Ingalls had received first samples from three other men, and these he personally took to the Arsenal. The first prism had its roof angle out of tolerance. The second followed the first into the rejection pile. But the third was different. Its angle was good, and the prism gave good definition in its image. Faith was restored, and with it came a trial order for fifty prisms.

Soon all three men succeeded in making good prisms and got small contracts for more. Then other members of The Gang were successful, and the amateur effort was established as a reliable source of roof prisms for the Arsenal. Now the men threw all their effort into the project. They gave up their spare hours and sometimes employed neighbors or members of the family. Some made ninety-eight per cent acceptance and better, compared to eighty- to-ninety per cent for most professionals. No one was more pleased or surprised at this record than the amateurs themselves.

Each man probably had his own reasons for making the prisms. They were, of course, paid for their products. However, there was much more to it than financial reward. For some it was a chance to prove that amateurs could produce precision optical parts as well as the professionals. For others it was a challenge that had to be met, an opportunity to learn a trade for peacetime, or a way of helping to win the war.

The Gang never grew large; about two dozen members with a few hired hands succeeded in contributing the total production of prisms. By 1943 the professional supply had about caught up with the demand and the need for the amateurs dwindled. In November 1944 Ingalls reported in *Scientific American* the results of the program he, Porter, and Ferson had worked on so long and hard to organize: The total prism production was over 28,000 accepted. Athough this figure was small compared to the professionals', perhaps ten per cent, Ingalls emphasized the fact that the acceptance records often exceeded those of the professionals. They even ran as high as ninety-nine and one-half per cent for the staggering 11,000 prisms produced by Ferson and his associate Paul Linde.

Sometime during the war Porter managed to find time to organize a commercial illustrating firm called "Russell W. Porter and Associates." His plan was to offer to industry his services as a graphic illustrator of complex to simple mechanisms. The associates were Porter's friends at Cal Tech. Their sales brochure was generously illustrated with the best of Porter's drawings of the 200-inch telescope and the observatory. They hoped that some of the companies working on war projects would be convinced that much time and money could be saved by employing three-dimensional cutaway drawings and other graphic techniques. Little is known of this venture except that it was rather short lived. Regrettably most of the men died before the business could become well established. But once again Russell Porter, though in his seventies, demonstrated his life-long willingness to begin a new venture.

XVI. THE FINAL YEARS

At the annual Stellafane conventions in 1940 and 1941, Porter had been the center of attraction as usual. Then they were discontinued because of the war. So in 1946, when they were revived, the Porters were eager to return to New England to see their old friends and home. One of the most exciting events of this trip occurred not at Port Clyde or in Springfield, but farther north. Porter had been invited to attend the commencement ceremony at Norwich University in Northfield, Vermont, where he had been a student for one year in 1889–1890. Now on June 9, 1946, he returned to the campus as one of five men to receive honorary degrees. Another on hand to accept a similar honor that day was Dwight David Eisenhower. The citation for the degree of Doctor of Science summed up Porter's accomplishments:

> Russell William (s) Porter—A cadet at Norwich in 1890; arctic adventurer and scientist who contributed photographic records, collected specimens and surveyed maps of hitherto unknown regions of the Far North, mechanical genius in the planning and construction of telescopes and mirrors; founder of the Amateur Telescope Makers Association, under whose influence and guidance thousands of people of all ages and in all parts of the world have been enabled to enjoy the marvelous beauties of the siderial universe with instruments of their own handiwork; recognized leader in the most ambi-

tious observatory project in all history; whose poetic skill with pencil and crayon has given structural beams and trusses a glamour appropriate to the 'music of the spheres' which they will help to interpret. Architect by training, mechanical designer, inventor and builder by profession, artist by native ability, giver of avocational enjoyment to numberless common people of whom he is proud to be one, living example of the truth that the most effective instructor of human beings is himself most human—Norwich University takes pleasure in conferring upon you the degree Doctor of Science.[1]

After the ceremony and receptions the Porters drove to Portland, Maine, to visit their daughter, who had been living there during the war. Years later Caroline enjoyed talking about this visit. While her parents were there the *Missouri* also visited the city and anchored in Casco Bay. An open house was being held for the people of Portland, a sea-oriented community, and Russell and Caroline decided to visit the famous battleship on which the Japanese surrender was signed. Porter carried on board a portfolio of photograhs of his war sketches. When they got on deck they discovered certain areas roped off and guarded by strategically stationed seamen. Russell and Caroline managed to get past a seaman and a rope and headed for a group of officers. Porter marched straight up to the Commander and introduced himself. From his portfolio he pulled a handful of photographs, and the next thing Caroline knew the three of them were in the officers' wardroom poring over the photos. Then the officer took them out to inspect the gun turrets, whose bearings were of particular interest to Porter. By the time the two men concluded their exchanges all the other visitors had left, and Russell and Caroline had to be taken back to shore in the officers' launch.

After leaving Portland, the first stop was down the coast of Maine at their home in Port Clyde, where they spent several weeks. Then about July 20, they drove to Springfield and stayed in the Hartness House, which after the death of Hartness in 1934 had been turned into an inn to cater to the businessmen visiting the machine tool companies in town. On August 3 Porter once again joined his telescope friends from many states and Canada on the hilltop at Stellafane. Four days of rain had threatened the weekend, but clearing weather gave them the chance to camp, catch up on the latest schemes for making telescopes, and exchange ideas among friends new and old. Over 375 people attended, a testimony to an undiminished desire to be TNs.

Porter was the first speaker on the program, which included numerous professional astronomers. He brought everyone up to date on the progress of the 200-inch telescope. The war had caused a delay in its completion, but now he expected it to be ready for use in about a year. He did not overlook his efforts for the war and told his admirers about his drawings of rocket intricacies and Japanese balloon bombs. Porter was followed by Drs. Donald Menzel and James Baker of Harvard Observatory, Professor George Dimitroff from Dartmouth College, Dr. John Strong of Johns Hopkins University, and Dr. Charles Smiley of Brown University. Dr. Menzel showed colored movies of solar prominences, taken at the Climax, Colorado, observatory. The time-lapse movies showed the gases exploding from the sun's surface in great streamers and arches thousands of times the volume of the earth.

On Monday after the weekend convention the Porters started their long journey back to California. This was the last time that Russell ever saw Stellafane. He was now seventy-four and had finished most of his work on the Palomar project. Although officially retired, "I still retain my office and putter around." As it had been for many years, his office was "the mecca of many transient mirror victims, a surprising number of Army and Navy fellows who had been bitten by the mirror microbe, who seem to like to drop in and talk shop for a while."[2]

In early 1947, to pursue his never-ending desire to work with optical glass, Porter decided to build a polishing machine and make some optical flats. He ended up making three eight-inch Pyrex flats good to one-tenth wave except at the edges, which were turned down about one-half wave, not quite good enough at the edges for high quality optics. He also made an eight-inch spherical mirror of ten-foot radius and used it with the Foucault knife-edge to test the flats.

Even though Porter enjoyed working on telescope mirrors and always considered himself an amateur when doing so, his supreme reward came while he watched the final steps in the fabrication of the 200-inch mirror and was one of the first men to see starlight brought to focus by the giant mirror, starlight that had been traveling through space since before the mirror had been conceived. Like

*Three Springfield Telescope Makers standing at the prime focus of the
nearly completed 200-inch telescope, March 1947*
Left to right: Oscar Marshall, Russell Porter, Ernest Flanders

most amateur mirrors that require a hole in the center for a Casse-grain telescope, the 200-inch mirror was ground and polished with a glass plug in its hole. In October 1947, with the grinding and polishing completed, the forty-inch stopgap was ready to be removed. This delicate operation required the plaster between the glass pieces to be chipped out by hand without damaging the mirror, and Porter was on hand to watch. He made a few sketches of the workmen chipping away at the plaster and shared them with the amateurs through *Scientific American.*

The mirror was then crated, hauled to the top of Palomar Mountain, and placed in the telescope. On the night of December 21, Dr. Anderson was the first man to look at a star with the new telescope. Porter and the other men who had been with the project from the beginning anxiously waited their turn. Each man realized that the mirror had not yet been figured to perfection near its edge and the stars would not show perfect images. But even so, after waiting nearly twenty years, it must have been an exciting and emotional experience to view the entire imaged star field suspended in space at the prime focal plane.

The mirror was then removed from the telescope and the final figuring was undertaken there in the obesrvatory. In early 1948 the first knife-edge tests were performed, using a star as the light source. Again Porter took part in the testing. Returning to Pasadena from the mountain, he dashed off a letter to Ingalls, which was printed in *Scientific American.* He told Ingalls that they had pointed the telescope at a ninth magnitude star (invisible to the naked eye) and that it had appeared so bright he had the feeling he could take it in his fingers and feel it. This was the first time Porter had ever seen the knife-edge test on a large telescope mirror using a star as the source. The atmospheric turbulence in front of the mirror reminded him of reeds or eel grass weaving back and forth in a rapidly flowing river.

Later in the night when the atmosphere calmed down and the seeing improved, the starlight was extinguished almost instantaneously over the entire mirror as the knife-edge cut the star's image. Next the men examined the star's image with an eyepiece, and it enlarged to nearly perfect circles inside and outside of focus. They were elated over the mirror and its support system even though more refinements were needed.

Russell W. Porter with a model of a universal telescope mount

Stellafane, Springfield, Vermont, as it appears today

Many of the men working on the 200-inch project, including Hale, had not lived to see it completed. Porter was more fortunate. He saw it finished and he saw it begin its long life of exploration of the universe, although illness kept him from its dedication on June 3, 1948.

Later in the summer of 1948 Porter decided that Alice and he could not take the trip east that year. Because he could not be with his friends at the annual convention, he substituted a letter. "You may be sure I regret not being present with you tonight, but the stress of other work has made this impossible. But I am very much with you in spirit." He told the amateurs that final adjustments on the mirror supports and final optical tests would be required before the 200-inch could begin doing actual work. In the meantime he had a chance to do some more observing. "I was looking at the Hercules Cluster the other night at the coude (f/30) focus with a power of about 900." The stars at the center of the cluster had never been resolved. "But they were completely separated against a dark sky background, and so bright as to require only a few seconds photographic exposure."[3]

Another project that involved Porter was the design of a new planetarium for the California Academy of Sciences at the Golden Gate Park in San Francisco. He must have enjoyed his work on the Griffith Panetarium in Los Angeles some years earlier, for now he accepted the new challenge, although only as a consultant. He went to San Francisco to advise on the early planning stages of the star projector. After a week of talking and thinking, Porter came up with a small sketch for the projector. It was instrumental in convincing the Academy Trustees that they had the skilled personnel and equipment to build the projector in their own shop. Although he was retained by the Trustees as a consultant, he did not see his scheme completed.

That same year Porter received the Amateur Astronomers Medal from the Amateur Astronomers Association of New York City. It was presented for his design work on the 200-inch Hale telescope.

Then in January 1949 a letter came to Porter from Dr. Samuel S. Stratton, President of Middlebury College, Vermont. "It is my pleasure to convey to you the unanimous vote of the President and Fellows of Middlebury College to confer upon you the Honorary degree of Doctor of Science."[4] Porter put off answering it.

In early February, Oscar Marshall paid a call on Porter at his home and found him busy in the basement working on a six-inch achromatic objective. This was the last time Porter ever saw his close friend. On February 22, he was planning to work as usual on his various projects. He finally answered the letter from President Stratton: "after due consideration I feel now unable to give you a positive answer in the affirmative."[5] He thought he might get to New England about commencement time, but would not know for some time.

After writing the letter he went down to the basement to work on his six-inch objective. Later that day he discussed the project with Dr. Anderson, who happened to drop by. But a short time after Anderson left, in the middle of the afternoon, Porter was stricken with a heart attack. A second attack at eleven-thirty that evening took his life. Suddenly, at age seventy-seven, it was over, the life of one man who had found true happiness in his pursuits of adventure, art, and science; but it was not forgotten. Chester Silvernail, a friend and colleague, once wrote, "I am personally glad that he lived and did the things he did because he certainly pointed out by his example how to have a happy life and some of the right things to do in this life."[6]

It fell to Alice to personally carry her husband's ashes across the country to the family plot in Port Clyde. Porter's final resting place was next to his infant son.

Of all the words written by Russell Porter and of all the words written about him, those which best epitomize the quality of his life were penned by him to a friend: "Nothing gives me more satisfaction than realizing that I have helped towards giving thousands of people the pleasure of creating with their own hands a tool to unlock the wonders of the heavens."[7]

ACKNOWLEDGMENTS

This is the first full-length biography of Russell W. Porter. Of all the sources of information, two were indispensable for the successful completion of this ten-year project. First, Mrs. Caroline Porter Kier, Russell Porter's daughter, was extremely generous in granting me taped interviews and allowing me to examine and use materials from her father's papers, diaries, unpublished manuscripts, and photographs, in particular the unpublished, "Arctic Fever," which is currently being published for the first time. Second, the Springfield Telescope Makers granted me free access to the archives at Stellafane containing records, letters, and photographs of Porter and of the early days of telescope making.

Mrs. Erika S. Parmi, librarian of the Stefansson Collection, Dartmouth College, aided me in my search for arctic material. Dr. Judith Goodstein, archivist, Millikan Memorial Library, Cal Tech, sent me copies of letters and telegrams between Porter and George E. Hale. Mr. Jonathan E. Kern sent me copies of letters between Porter, Fred B. Ferson, and J. Adair Lyon. Mr. Robert J. Kieckhefer sent me material on the Japanese balloon bombs and the Arctic.

My special thanks are extended to those who granted me interviews and wrote me letters concerning Russell Porter. Their names are included in the Notes and Reference sections that follow.

Special gratitude is given to my wife, Sylvia, for critical reading of the manuscript, and to Mrs. Francis Roberts, my mother-in-law, and my late mother, Mrs. Vera D. Willard, for editing of the manuscript. I would like to express my sincere thanks also to Mr. David O. Woodbury for his critical reading of the manuscript and for his introduction.

Berton Willard

NOTES

Chapter 1

1. R. W. Porter, "Arctic Fever" (unpublished manuscript, 1933, Dartmouth College), p. 2.

2. Letter dated August 10, 1894, Sukkertoppen, Greenland.

3. Porter, "Arctic Fever," p. 6.

4. Ibid., p. 7.

5. Ibid., p. 10.

6. Ibid., p. 11.

7. Ibid., p. 13.

Chapter 2

1. R. W. Porter, "Arctic Fever" (unpublished manuscript, 1933, Dartmouth College), p. 20.

2. Ibid., p. 23.

3. Ibid., p. 24.

4. Ibid., p. 31.

5. Ibid., p. 33.

6. Ibid.

7. Ibid., p. 37.

8. Robert E. Peary, *Northward over the Great Ice* (New York: Frederick A. Stokes Company, 1898), I, xlix.

9. Porter, "Arctic Fever," p. 41.

10. Ibid., p. 42.

11. Ibid., p. 43.

12. Ibid., p. 44.

13. Ibid., p. 39.

Chapter 3

1. R. W. Porter, unpublished diary of the Ziegler Polar Expedition, 1903–1905 (Dartmouth College), p. 1.
2. Ibid., p. 2.
3. Ibid., p. 10.
4. Ibid., p. 24.
5. Ibid., p. 29.
6. R. W. Porter, "Arctic Fever" (unpublished manuscript, 1933, Dartmouth College), p. 52.

Chapter 4

1. R. W. Porter, unpublished diary of Ziegler Polar Expedition, 1903–1905 (Dartmouth College), p. 38.
2. Anthony Fiala, *Fighting the Polar Ice* (New York: Doubleday, Page and Company, 1907), p. 82.
3. R. W. Porter, "Arctic Fever" (unpublished manuscript, 1933, Dartmouth College), p. 56.
4. Fiala, *Fighting the Polar Ice*, p. 90.
5. Porter, unpublished diary, p. 46.

Chapter 5

1. R. W. Porter, unpublished diary of Ziegler Polar Expedition, 1903–1905 (Dartmouth College), p. 50.
2. Ibid., p. 51.
3. Ibid., p. 47.
4. Ibid., p. 52.
5. Ibid., p. 55.
6. Ibid., p. 82.
7. Ibid., p. 61.
8. Ibid., p. 65.
9. Ibid., p. 92.
10. Ibid., p. 99.
11. Ibid., p. 101.

Chapter 6

1. R. W. Porter, unpublished diary of Ziegler Polar Expedition, 1903–1905 (Dartmouth College), p. 110.
2. R. W. Porter, "Arctic Fever" (unpublished manuscript, 1933, Dartmouth College), p. 72.
3. Porter, unpublished diary, p. 120.

Chapter 7

1. Also spelled Traleika and Doleika.
2. Spelling is Cook's. Also spelled *Santa Anna* by Parker. See reference.
3. Cook shows a photo in his book with *Bolshoy* painted on the

launch. Porter wrote *Bolshoy* and Browne wrote *Bulshaia.* See references.

4. Cook wrote seven days, Browne wrote three. See references.

5. Belmore Browne, *The Conquest of Mount McKinley* (Boston: Houghton Mifflin Company, 1956), pp. 29–30.

6. Browne and Parker wrote one man, Porter and Cook wrote two. See references.

7. R. W. Porter, "Arctic Fever" (unpublished manuscript, 1933, Dartmouth College), p. 80.

8. Cook, in his book *My Attainment of the Pole,* claimed Barrill had been bribed into making the statement.

9. Porter, "Arctic Fever," p. 79.

Chapter 8

1. Joseph W. Roe, *James Hartness* (New York: The American Society of Mechanical Engineers, 1937), p. 114.

2. I have had the occasion to test the sixteen-inch plate glass mirrors and found the flat had a turned-down edge of about one and one-quarter wavelengths of light and the parabola about one-half wave, still remarkable considering the conditions in which they were made.

3. R. W. Porter, "An Amateur's Observatory," *Scientific American Supplement,* August 4, 1917, pp. 68–69.

4. R. W. Porter, "Moonscapes," *Popular Astronomy,* October 1916, p. 515.

5. Letter from James Hartness to Russell Porter, in collection of Caroline Porter Kier.

6. Letter of Alva H. Bennett to the author, April 11, 1967.

Chapter 9

1. Russell W. Porter, *Springfield Reporter,* March 4, 1920.

2. Ralph E. Flanders, taped interview with the author, December 11, 1965.

3. Ibid.

4. Ibid.

Chapter 10

1. Springfield Telescope Makers, Secretary's Records, Springfield, Vermont.

2. Ibid.

3. Oscar S. Marshall, *Springfield Reporter,* June 18, 1925, p. 1.

Chapter 11

1. *Springfield Reporter,* July 19, 1923.

Chapter 12

1. Carnegie Institute of Washington, Millikan Memorial Library, Hale letter file.

2. Ibid.

3. R. W. Porter, "Moonscapes," *Scientific American*, October 1930, pp. 256–257.

4. To learn the complete story, the reader is again referred to *The Glass Giant of Palomar*, by David O. Woodbury, and *Palomar: The World's Largest Telescope*, by Helen Wright.

5. Letter, R. W. Porter and O. S. Marshall to the Springfield Telescope Makers, January 1, 1932. Records of the Springfield Telescope Makers.

6. Letter, O. S. Marshall to the Springfield Telescope Makers, December 19, 1932. Records of the Springfield Telescope Makers.

7. Letter, R. W. Porter to members of the amateur telescope fraternity. Records of the Springfield Telescope Makers.

8. Letter, R. W. Porter to R. J. Lyon, January 20, 1935. Records of the Springfield Telescope Makers.

9. Records of the Springfield Telescope Makers.

Chapter 13

1. John C. Pierce. Reading, Vermont. Personal taped interview with the author. November 1, 1968.

2. Letter, Knowles Ryerson to the author, February 11, 1969.

3. R. W. Porter, "Sun Dials and Sun Dialing," *Scientific American*, August 1928, pp. 150–152.

4. Ibid.

5. R. W. Porter, "Telescoptics," *Scientific American*, February 1938, pp. 120–121.

6. R. W. Porter, "Arctic Fever" (unpublished manuscript, 1933, Dartmouth College), p. 8.

7. Letter, R. W. Porter to Lillian G. Sievers, July 21, 1944.

8. Letter, Russell W. Porter and Oscar S. Marshall to the Springfield Telescope Makers. Records of the Springfield Telescope Makers.

9. Ibid.

10. Records of the Springfield Telescope Makers.

11. Letter, Russell W. Porter to George A. Perry, May 27, 1932. Records of the Springfield Telescope Makers.

12. Letter, Russell W. Porter to George Perry, March 12, 1933. *Springfield Reporter*.

13. Letter, Russell W. Porter to Roy J. Lyon, January 20, 1935. Records of the Springfield Telescope Makers.

Chapter 14

1. Letter, Russell W. Porter to Arthur D. Baker, December 10, 1932. Records of the Springfield Telescope Makers.

2. Letter, Russell W. Porter to the amateur telescope fraternity, July 16, 1934. Records of the Springfield Telescope Makers.

3. Letter, Russell W. Porter to George E. Hale, July 29, 1931. California Institute of Technology, Millikan Memorial Library, Hale letter file.

4. Letter, Russell W. Porter to George Perry, March 12, 1933. *Springfield Reporter.*

5. Ibid.

6. Ibid.

Chapter 15

1. Letter, Oscar S. Marshall to the Springfield Telescope Makers, April 28, 1944. Records of the Springfield Telescope Makers.

2. Albert G. Ingalls, "A Hobby Goes to War," *Scientific American,* May 1943, pp. 202–205.

3. Ibid., p. 203.

4. Ibid.

Chapter 16

1. *Norwich University Record,* October 3, 1947.

2. Letter, Russell W. Porter to Leo Scanlon, November 3, 1946. Leo Scanlon.

3. Letter, Russell W. Porter to Stellafane. Printed in *Sky and Telescope,* October 1948, p. 297.

4. Mrs. Caroline Porter Kier.

5. Ibid.

6. Letter, Chester J. Silvernail to the author, September 12, 1971.

7. Letter, Russell W. Porter to Leo Scanlon, November 3, 1946.

REFERENCES

Anderson, J. A., and R. W. Porter. "The 200-Inch Telescope," *The Tele-scope*, March-April 1940, pp. 29–39.

Anderson, J. A. "The Astrophysical Observatory of The California Institute of Technology." *Journal of the Royal Astronomical Society of Canada*, Vol. 36, May-June 1942, pp. 177–200.

Babcock, Horace W. Hale Observatories. Letter to the author, September 23, 1970.

Baker, Mary E. *Folklore of Springfield*. Springfield, Vermont: The Altrurian Club of Springfield, Vermont, 1922.

Bennett, Alva H. Thompson, Connecticut. Letter to the author, April 11, 1967.

Broehl, Wayne G., Jr. *Precision Valley*. Englewood Cliffs, New Jersey: Prentice-Hall Inc., 1959.

Browne, Belmore. *The Conquest of Mount McKinley*. Boston: Houghton Mifflin Co., 1956.

Bryant, Henry G. "An Expedition in South-Eastern Labrador," *The Geographical Journal*, Vol. 61, April 1913, pp. 340–346.

California Institute of Technology. Millikan Memorial Library, Hale letter on file.

Choukas, Michael Jr., Headmaster, Vermont Academy. Letter to the author, April 11, 1966.

Collinso, Richard Rear-Admiral. *The Three Voyages of Martin Frobisher*. New York: Burt Franklin, Publisher, undated. Reprinted from the first edition of *Hakluyt's Voyages*, 1867.

Cook, Frederick A. *To the Top of the Continent*. New York: Doubleday, Page and Company, 1908.

———. *My Attainment of the Pole*. New York: Mitchell Kennerley, 1913.

Ferson, Fred B. "Prisms, Flats, Mirrors," *Amateur Telescope Making Advanced*. Edited by Albert G. Ingalls. New York: Scientific American Inc., 1954.

Fiala, Anthony. *Fighting the Polar Ice*. New York: Doubleday, Page and Company, 1907.

Flanders, Ralph E. Springfield, Vermont. Personal taped interview with the author, December 11, 1965.

Fleming, John A., ed. *The Ziegler Polar Expedition, 1903–1905, Scientific Results*. Washington, D.C.: National Geographic Society, 1907.

Flinn, Henry. Springfield, Vermont. Personal interview with the author, July 23, 1968.

A Forty-Five Year History of the Class of 1896. Massachusetts Institute of Technology, 1942.

Gourdeau, B. Sierra Madre, California. Letter to E. Washburn, Springfield, Vermont, March 1968. E. Washburn.

Hartness, James. Letters to R. W. Porter, 1913–1917. Collection of Caroline Porter Kier.

The Hartness Screw Thread Camparator. Springfield, Vermont: Jones and Lamson Machine Company, 1921.

Hayward, Roger. Pasadena, California. Letter to the author, May 8, 1970.

Hill, John C. II Commander U. S. Navy; Thomas Utegaard, F. Lt. Commander U.S. Navy; and Gerard Riordan. *Dutton's Navigation and Piloting*. Annapolis: United States Naval Institute, July 1967.

History of the Town of Springfield, Vermont. Boston: Walker and Co., 1896.

Ingalls, Albert G. "The Heavens Declare the Glory of God," *Scientific American*, November 1925, pp. 293–295.

————. *Amateur Telescope Making*. N. P.: Scientific American Publishing Company, 1926.

————. "The Telescope Making Enthusiasts Convene," *The Scientific American Digest*, October 1927, pp. 345–346.

————. "California Amateurs Make Telescopes," *Scientific American*, March 1928, pp. 244–245.

————. "The Back Yard Astronomer," *Scientific American*, May 1928, p. 448.

————. "The Amateur Astronomer," *Scientific American*, June 1934, pp. 324–325.

————. "The Amateur Astronomer," *Scientific American*, July 1934, pp. 44–45.

————. "The Amateur Astronomer," *Scientific American*, June, 1935, pp.318–320.

————. "The Amateur Telescope Maker," *Scientific American*, November 1935, pp. 276–278.

————. "The Amateur Telescope Maker," *Scientific American*, January 1937, pp. 42–45.

————. "Telescoptics," *Scientific American*, July 1939, pp. 58–61.

———. "Telescoptics," *Scientific American*, August 1940, pp. 106–109.

———. "Telescoptics," *Scientific American*, February 1941, pp. 122–125.

———. "Telescoptics," *Scientific American*, October 1941, p. 234.

———. "Making Prisms," *Scientific American*, May 1943, pp. 202–205.

———. "Telescoptics," *Scientific American*, May 1943, pp. 238–239.

———. "Telescoptics," *Scientific American*, November 1944, pp. 239–240.

———. "Telescoptics," *Scientific American*, June 1946, pp. 285–286.

———. "Telescoptics," *Scientific American*, March 1947, pp. 143–144.

———. "Telescoptics," *Scientific American*, January 1948, pp. 45–48.

———. "Telescoptics," *Scientific American*, March 1948, pp. 141–144.

———. "The Amateur Astronomer," *Scientific American*, April 1949, pp. 60–63.

———. "The Amateur Astronomer," *Scientific American*, May 1949, pp. 60–63.

———. "The Amateur Astronomer," *Scientific American*, August 1949, pp. 60–63.

———. ed. *Amateur Telescope Making* Books One, Two, and Three. New York: Scientific American, Inc., Book One 1952, Book Two 1954, Book Three 1956.

Ingalls, Mrs. Albert G. Letter to the author, February 4, 1968.

Ingalls, Jeremy G. Nutley, New Jersey. Letter to the author, April 17, 1967.

Johnson, Joseph B. Springfield, Vermont. Personal taped interview with the author, winter 1968.

Kaufmann, William J. Griffith Observatory and Planetarium. Letter to the author, May 7, 1970.

Kern, Jonathan E. New Orleans, Louisiana. Letters between Russell W. Porter, Fred B. Ferson, and J. Adair Lyon, 1939–1941.

Kier, Mrs. Caroline Porter (daughter of Russell W. Porter). Personal taped interviews with the author, 1966–1971.

King, Henry C. *The History of the Telescope*. Cambridge, Massachusetts: Sky Publishing Corporation, 1955.

Loudon, Henry A. Rye Beach, New Hampshire. Letter to the author, April 26, 1968.

"Lunar Crater Porter," *Sky and Telescope*, June 1949, p. 206.

Mackintosh, Alan. Bloomington, Indiana. Letter to the author, September 14, 1972, and other personal communications.

Marshall, Oscar S. "Russell W. Porter." *The Vermonter, The State Magazine*, Vol. 33, no. 8, 1928, pp. 118–122.

———. "Russell W. Porter As I Knew the Man," *Springfield Reporter*, May 11, 1949.

———. "Russell W. Porter 1871–1949," *Popular Astronomy*, May 1949, pp. 235–40.

———. Letters to the Springfield Telescope Makers. Records of the Springfield Telescope Makers.

Marshall, Oscar S. Unpublished autobiography. Mrs. Beulah Wright, daughter of O. S. Marshall.

Marvin, Esther W. Secretary to the Vice-President of the Alumni Association, Massachusetts Institute of Technology. Letter to the author, April 20, 1966.

Mayall, R. Newton, and Margaret L. Mayall. *Sundials.* Boston: Hale, Cushman and Flint, 1938.

Mazuzan, John E. Alumni Secretary, Norwich University. Letter to the author, April 23, 1966.

Miller, Robert C. Letters to the author, October 25 and December 19, 1974

Miller, William C. Hale Observatories. Letter to the author, January 23, 1973.

Moore, Terris. *Mount McKinley: The Pioneer Climbs.* College, Alaska: University of Alaska Press, 1967.

New York Herald Tribune. "Dean of Amateur Star Gazers." August 28, 1932.

Norwich University Record, Vol. 39, no. 5, October 3, 1947.

Parker, Herschel C. "The Exploration of Mt. McKinley: Is It the 'Crest of the Continent'?" *The American Monthly Review of Reviews,* January 1907, pp. 49–58.

Peary, Robert E. *Northward Over the Great Ice.* New York: Frederick A. Stokes Company, 1898.

———. *The North Pole.* New York: Frederick A. Stokes Company, 1910.

Perkin, Richard S. Norwalk, Connecticut. Letter to the author, March 26, 1969.

Pierce, John C. Reading, Vermont. Personal taped interview with the author, November 1, 1968.

Popolo, D. Rotch Library, Massachusetts Institute of Technology. Personal communication with the author, September 22, 1972.

Porter, Russell W. Unpublished diary of Ziegler Polar Expedition, 1903–1905, Dartmouth College.

———. "Moonscapes," *Popular Astronomy,* October 1916, pp. 515–516.

———. "A New Projection of the Moon," *Popular Astronomy,* November 1916, pp. 573–574.

———. "The Enclosed Observing Room," *Popular Astronomy,* May 1917, pp. 296–300.

———. "An Amateur's Observatory," *Scientific American Supplement,* August 4, 1917, pp. 68–69.

———. "A New Form of Mounting for Large Reflectors," *Popular Astronomy,* March 1918, pp. 147–179.

———. "Knife-Edge Shadows: Photography as an Aid in Testing Mirrors," *The Astrophysical Journal,* Vol. 47, June 1918, pp. 324–328.

———. "Latitude without Instruments," *Popular Astronomy,* April 1921, p. 197.

———. "The Telescope Makers of Springfield, Vermont," *Popular Astronomy*, March 1923, pp. 153–162.

———. "The Garden Telescope," *House Beautiful*, March 1923.

———. "A Garden Telescope," *Popular Astronomy*, May 1924, pp. 273–280.

———. "Mirror Making for Reflecting Telescopes," *Scientific American*, February 1926, pp. 86–89.

———. "Mountings for Reflecting Telescopes," *Scientific American*, March 1926, pp. 164–167.

———. "Stellafane," *Popular Astronomy*, November 1927, pp. 501–505.

———. "Sun Dials and Sun Dialing," *Scientific American*, August 1928, pp. 150–152.

———. "Arctic Fever." Unpublished manuscript, 1933, Dartmouth College.

———. "The Telescope Craze Gains Favor." *The Telescope*, October 1933, pp. 14–23.

———. "Sun Clocks," *Scientific American*, August 1935, pp. 84–85.

———. "Telescoptics," *Scientific American*, February 1938, pp. 120–121.

———. Letter to Lillian G. Sievers, Washington, D.C. July 21, 1944. Lillian G. Sievers.

———. Letter to Leo Scanlon, November 3, 1946. Leo Scanlon, Pittsburgh, Pennsylvania.

———. "Autobiographical Notes." c. 1946, Dartmouth College.

———. "Moonscapes," *Scientific American*, October 1930, pp. 256–257.

———. "Telescoptics," *Scientific American*, May 1948, p. 63.

———. "R. W. Porter's Letter to Stellafane," *Sky and Telescope*, October 1948, p. 297.

———. Letters to the Springfield Telescope Makers. Records of the Springfield Telescope Makers.

——— and Oscar S. Marshall. Letters to the Springfield Telescope Makers. Records of the Springfield Telescope Makers.

The Porter Garden Telescope. Springfield, Vermont: Jones and Lamson Machine Company, 1923.

Portland Evening Express. June-July 1946.

Reyerson, Knowles A. "Russell W. Porter—An Appreciation," *Popular Astronomy*, November 1949, pp. 436–439.

———. Davis, California. Letters to the author, May 15, 1968, February 11, 1969.

Rockland Gazette, Rockland, Maine, 1907.

Roe, Joseph W. *James Hartness.* New York: The American Society of Mechanical Engineers, 1937.

Rule, Bruce H. Hale Observatories. Letter to the author, March 12, 1973.

Scanlon, Leo J. "The Porter Garden Telescope," *Popular Astronomy*, January-February 1962, pp. 22–25.

Siever, Lillian G. Letters to the author, May 26, July 9, August 7, 1969.

Silvernail, Chester J. Los Alamos, New Mexico. Letter to the author, September 12, 1971.

Springfield Reporter. Springfield, Vermont, 1905–1907; 1919–1946.

The Springfield Telescope Makers. Secretary's reports and other records, Springfield, Vermont.

"Stargazers at War," *Time.* October 4, 1943, pp. 86–88.

Stokley, James. "He Showed Thousands the Stars," *Science Newsletter,* December 7, 1929, pp. 349–351.

Strong, John. Amherst, Massachusetts. Personal taped interview with the author, November 3, 1968.

Town of Springfield, Vermont. Records.

Waldron, Webb, "Stars on a Mountain." *Century Magazine,* April 1925, pp.786–793.

———. "One Really Happy Man." *The American Magazine,* November 1931, p. 50.

———. *Americans.* New York: The Greystone Press, 1941.

Walsh, Henry Collins *The Last Cruise of the Miranda,* New York: The Transatlantic Publishing Company, 1895.

Washburn, Ethel. Letters from Bessie Gourdeau, received by Miss Washburn in March 1968, given to the author.

Welch, G. B. Frankford Arsenal. Letter to Russell W. Porter, June 12, 1943.

Wheeler, Merric. Springfield, Vermont. Personal interview with the author, July 22, 1968.

Whitaker, Ewen A. Lunar and Planetary Laboratory, Tucson, Arizona. Letter to the author, September 22, 1969.

Woodbury, David O. *The Glass Giant of Palomar.* New York: Dodd, Mead and Company, 1963.

Wright, Helen. *Palomar: The World's Largest Telescope.* New York: The MacMillan Company, 1952.

Zolotas, Constance S. Alumni Office, The University of Vermont. Letter to the author, April 22, 1966.